D.

# MATHEMATICAL MYSTERIES

## The Beauty and Magic
of Numbers

# MATHEMATICAL MYSTERIES

## The Beauty and Magic of Numbers

CALVIN C. CLAWSON

PLENUM PRESS • NEW YORK AND LONDON

Library of Congress Cataloging-in-Publication Data

---

Clawson, Calvin C.
    Mathematical mysteries : the beauty and magic of numbers / Calvin
C. Clawson.
        p.    cm.
    Includes bibliographical references (p.    -   ) and index.
    ISBN 0-306-45404-1
    1. Number theory.    I. Title.
QA241.C664   1996
512'.7--dc20                                                    96-31715
                                                                    CIP

---

ISBN 0-306-45404-1

© 1996 Calvin C. Clawson
Plenum Press is a Division of Plenum Publishing Corporation
233 Spring Street, New York, N.Y. 10013-1578

10 9 8 7 6 5 4 3 2 1

Printed in the United States of America

To my best friend and wife, Susan

# CONTENTS

# ACKNOWLEDGMENTS

Many have generously given of their time to make this book possible. Special thanks go to my workshop friends who patiently reviewed the manuscript and offered valuable suggestions: Marie Edwards, Bruce Taylor, Linda Shepherd, Phyllis Lambert, and Brian Herbert. A very special thanks goes to Martin Gardner for allowing me to snoop through his files on prime numbers. I also want to thank Larry Curnutt for inspiring the two chapters on Ramanujan and for reviewing the rough draft, providing many excellent recommendations. Thanks go to David Bennett for reviewing the manuscript and identifying several difficulties. I cannot fail to mention my indebtedness to the mathematics and philosophy professors at the University of Utah, who were both kind and inspiring while leading me from the darkness into the light.

# INTRODUCTION

*The mathematician lives long and lives young; the*

*wings of the soul do not early drop off, nor do its*

*pores become clogged with the earthly particles blown*

*from the dusty highways of vulgar life.*

JAMES JOSEPH SYLVESTER

(1814–1897)[1]

*I* vividly remember a class on numerical analysis taught by Professor Chamberlain at the University of Utah during the 1960s. He would become so enthralled with his lecture that, while hurriedly writing equations on the blackboard, he would fail to notice how closely he was stepping to the edge of the platform, raised six inches off the floor. My attention fluctuated between following his lecture and watching his feet move ever nearer to the platform's edge. Suddenly and without warning, he would step off the platform and fall in a great tumble to the floor.

The entire class would burst out laughing. Chamberlain would give a great laugh as he picked himself off the floor and brushed the dust from his pants. Smiling, he would offer some delightful joke and then return to the blackboard and his equations. We were so awed by his complete involvement in the realms of numerical analysis that we, too, paid intense attention to his lectures trying to discover what he saw in the subject matter. By the end of the year, we had fallen a little more in love with numbers.

1

But why, we might ask, should we study mathematics at all? Most adults admit to an ignorance and a deeply rooted annoyance of mathematics. Why should we then torture ourselves by studying the subject when we are not required to do so? On the first day of class I ask my algebra students why anyone should pursue mathematics. They never fail to offer a list of valid reasons: mathematics is required to finish their schooling; it helps them in their personal finances; it will be required on the job; it helps them tutor their children; it promotes their understanding of science. All these are good reasons, each sufficient in itself. Yet each of these reasons misses the primary motivation to study mathematics.

The single most compelling reason to explore the world of mathematics is that it is beautiful, and pondering its intriguing ideas is great fun. I'm constantly perplexed by how many people do not believe this, yet over 50,000 professional mathematicians in America practice their trade with enthusiasm and fervor. Another five to ten million Americans study mathematics for the pure joy of it, without any anticipation of ulterior rewards. Can all these people be wrong? Certainly not! They have all learned a great secret: to study the deep truths of number relationships feeds the spirit as surely as any of the other higher human activities of art, music, or literature.

Even though I state with conviction that mathematics is fun, those people suffering from math phobia or experiencing math revulsion will continue to shake their heads, failing to take that first step forward to look into the mathematical depths. When they open a book on mathematics and encounter all the strange symbols strewn across the pages, their stomachs tie into knots and their hearts begin racing. Mathematics is simply not for them. This too-common reaction is a conditioned response programmed into their psyche from childhood.

Now, for a confession. Even though I've received a college degree in mathematics, and have published books on the subject, I, too, still occasionally suffer from math anxiety. At the local library or bookstore I might find a math book with an interesting title. Flipping through the pages, I'm struck with numerous bizarre symbols which I've not seen before, and of which I have no understanding. My stomach takes a hop, and my heart races. My discomfort at that moment is the same as that of the math phobic. The difference is that I know the feeling of discomfort will pass, that it is only an automatic response of my body when my mind is faced with an unknown situation. I also know that if I choose to purchase the book and learn the definitions involved, I'll begin to harvest a great enjoyment from the material. Therefore, it is possible for all who suffer math phobia and math anxiety to reverse the situation and learn to enjoy mathematics. Enjoyment is the keystone for this volume. It is written to present to you, the reader, fundamental mathematical truths to dazzle and amaze you.

The greatest discovery of all humankind may well have been the natural numbers. The natural or "counting" numbers have been with us since prehistoric times, assisting us in our struggle to emerge from the primitive lifestyle of the hunter–gatherer to become the modern human beings of the 21st century. For many past aeons, philosophers and mathematicians have studied the sequence of natural numbers, uncovering startling and mystifying truths. Mathematics, itself, began with the natural numbers and the study of their relationships. The demand for solutions to new and sophisticated problems encountered in our march from simple farmers to merchants, priests, scientists, and finally modern industrialists, forced us to use the natural numbers to construct the great edifice of modern mathematics.

Mathematicians, in their search for these solutions, progressed beyond the natural numbers, discovering fractions, irrational, transcendental, transfinite, and surreal numbers. However, the majority of professional mathematicians recognize that the most important problems in mathematics today still involve the natural number sequence.

By studying the natural numbers in their wonderful ordered sequence, we discover a perplexing and utterly charming characteristic of mathematics—its interconnectedness. The reasons why seemingly unrelated mathematical truths are connected in simple and beautiful equations continue to stump mathematicians. The natural number sequence also represents our first encounter with the idea of infinity, an idea that still puzzles and entertains us. Finally, the very notion of numbers forces us to deal with the question of what reality is. Do numbers exist without humans to think of them? If so, what exactly are these "ideas" and how is their existence different from the material world we see around us?

Frequently, when studying mathematics, one slips into another world, a world of exquisite beauty and truth. This traveling to another plane of mental existence can be so addicting that the practitioner is lost, like Professor Chamberlain, to ordinary, daily stimulus.

Because this book is about mathematics, you will find formulas and equations on some of the pages. A current trend is under way to write books about mathematics and science that do not contain any, or at least very few, equations. My own editor said that her publishing house would like a book without all those pages and pages of equations. In the foreword to Stephen Hawking's book, *A Brief History of Time: A Reader's Companion*, Professor Hawking ruminated about advice he has received regarding the writing of his popular book.

> Each equation, I was told, would halve the sales of the book. But that was okay. Equations are necessary if you are doing account-ancy, but they are the boring part of mathematics. Most of the interesting ideas can be conveyed by words or pictures.[2]

However, to ignore equations in our study of mathematics is wrong. Learning about mathematics without equations is like eating chocolate candies without the delicious, gooey centers. They can temporarily satisfy, but we miss the yummy and mysterious candied sweets in the middle. To push this analogy even further, equations are much like chocolates. Equations are small and compact, while hiding delicious interiors of intriguing truths. A simple example will illustrate.

For most of my adult life, if anyone had shown me the equation $X^3 + 1 = 0$ and asked what I thought, I surely would have yawned and shaken my head in boredom. The equation appears to have no special features to draw our attention. Then I discovered the wonderful sweets hidden in the center of this chocolate. The equation has three solutions for $X$—values for $X$ that make the lefthand side of the equation 0. One obvious solution is when $X$ is equal to $-1$. That is: $(-1)^3 + 1 = -1 + 1 = 0$. However, two other solutions exist. These solutions are not examples of ordinary numbers (real numbers) we find on a number line, but are examples of the strange complex numbers. The complex numbers occupy an entire plane rather than just a line. A popular method of solving such equations is called Newton's Method. If we use this method on $X^3 + 1 = 0$ and graph the results, we discover a complicated, yet beautiful image that replicates itself under magnification. Hence, when we magnify the image it never resolves itself into simple lines and curves, but always remains complex. Such images are called fractal by mathematicians. Figure 1 is the fractal image associated with the equation $X^3 + 1 = 0$. Why this image is associated with this particular equation, nobody can say—it's a mystery. Now, when encountering this little equation, a shock runs up my spine, for I understand that hidden under its apparent simplicity and elegance is a deep and complex mystery which we have still to solve. Words alone cannot convey the essence of this equation or feed my hunger for its delectable center: I need the tasty little equation, itself.

Therefore, we will look at interesting equations, even though we are forewarned that each may "halve the sales of the book." But eating too many chocolates too quickly can cause an upset stomach. So we'll be judicious, and pace our equations. In this way, each one we encounter will seem like a special treat.

Before beginning our journey I must add one more note. Books, including books about mathematics, are changing in a fundamental way: Many of the facts and opinions reported within books are now available to all those who own computers. Because we are in the very nucleus of this change, it is impossible to predict just what shape things will take in the future. This situation has occurred because of the Internet, the great sharing of ideas through the millions of connected computers around the world. For example, it is now possible to dial into the Internet and discuss any kind of mathematical question (or any other kind of question for that

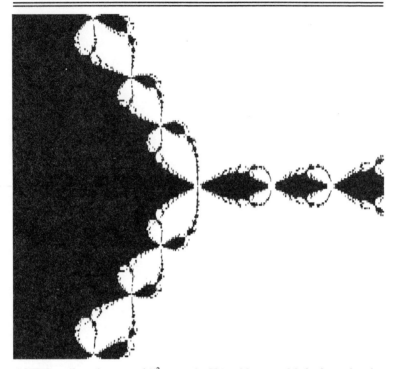

FIGURE 1. Fractal image of $X^3 + 1 = 0$. Using Newton's Method to solve the equation $X^3 + 1 = 0$, we discover that the initial values used to find the roots of the equation in the complex plane do not always lead to the closest root, but generate this fractal.

matter) with professional and amateur mathematicians all over the world.

Not long ago I needed information that I could not find at my local library or from my collection of math books. I left a request for the information on the Internet. Within hours I received my first reply. Over the next few days I received a half-dozen E-mail responses, giving me many useful references on the subject. Responses came from such diverse sources as a professional mathematician in England and a graduate student in California. One response even gave the name of a book which I already had in my own bookcase, but had failed to realize contained a good discussion of my problem.

The various computers connected through the Internet maintain files of information. By using the Internet (called "surfing the net" by aficionados) it is possible to look into those files that are open to the public. Such public files can be repositories of intriguing bodies of knowledge. At one computer site is a file containing the biographies of 350 mathematicians.[3] Another location contains a list of all the largest prime numbers.[4] A third file[5] contains the first 1.25 million digits of $\pi$.

Just a few years ago, it was almost impossible to find such current information, unless you were very familiar with the specific mathematical field involved. Today, through the Internet, this information is freely available to the general public. Some of the references I cite come from the Internet, especially in areas where knowledge is changing so fast that it has not filtered into book form, or that books, as a medium of information transfer, are simply too slow to record. This great diversity and depth of knowledge can be at the fingertips of anyone with access to a computer and a modem. Today the amateur mathematician can communicate with the most brilliant professional, sharing both ideas and the pleasure of probing the great mysteries of the cosmos. Just how the Internet will change our current methods of disseminating knowledge we can't say.

Whatever does happen, we can be certain of one thing: It will be exciting. Just as the invention of movable type gave us printing and the democratic spread of knowledge in the 15th century, the computer and the Internet will greatly accelerate the pace of our knowledge sharing in the 21st century. How this will change the face of mathematics is anyone's guess.

## CHAPTER ONE

# DISCOVERY OF THE NUMBER SEQUENCE

*I had been to school most of the time, and could spell,*

*and read, and write just a little, and could say the*

*multiplication table up to six times seven is thirty-*

*five, and I don't reckon I could ever get any further*

*than that if I was to live forever. I don't take no stock*

*in mathematics, anyway.—Huck Finn*

MARK TWAIN

THE ADVENTURES OF HUCKLEBERRY FINN[1]

## WHAT IS COUNTING?

*T*he natural number sequence is the list of counting numbers used in everyday life. They are such an important component of our general knowledge that they are one of the first things we teach our children. When our babies are two or three years old, we begin by holding their hands up and pointing to each successive finger as we say the appropriate word for that number of fingers. We are encouraged when our children say the correct number back to us: "one, two, three, four, . . ." But is this counting? Not quite.

We could actually teach our children another sequence of words which have no number meaning, such as the lyrics to a song. For example, we could teach them the sequence: up, up in the sky, where the little birds fly . . . Saying these words, even in the correct order, is not counting, for counting is carried out with a specific purpose in mind, a final goal. A list of words such as sky and fly, as the words of a song, are also learned in a specific order, but do not represent counting. Order is certainly necessary to counting, but simply saying words in order leaves out something very important. To count implies we are after a special end result to our activity. To understand what that goal may be we must look at what numbers are used for.

In general, numbers have three uses, and are given three names to reflect these uses. Cardinal numbers are used to find how many objects are in a collection. We call the objects "elements" and the collections "sets." This is the first and most important use of numbers.

Ordinal numbers are used to find the proper order of elements in a set. If you graduated number three in your college class, then you were after two other students. The "three" does not tell how many students graduated, but only where you stood among those graduates. Therefore, the "three" is used as an ordinal number but not a cardinal number. Hence, ordinal numbers give us information about an element's position within a set, without identifying the cardinal number of the set. There are many uses of ordinal numbers. For example, street addresses are ordinal numbers. An address of 1933 Spring Street does not tell us how many houses are on Spring Street or how many houses are in the city. But the number 1933 does tell us that this house comes before 1947 Spring Street and after 1909 Spring Street. Some numbers may be used as both cardinal and ordinal numbers, e.g., counting on our fingers while speaking the counting words. In this case we not only determine the total number of fingers (cardinal number), but we also assign each number to a specific finger, and, hence, determine an ordering for the fingers.

The last use of numbers is simply to give a unique name to an object. These are tag numbers. Phone numbers, airline flight numbers, and bus numbers are tag numbers. Tag numbers give us no information about the number of phone customers, airline flights, or bus routes, nor do they give us any information about the order of items in a set. We can't say flight 302 leaves before or after flight 1601. As numbers, tag numbers are of the least interest because they are really used in the place of proper names. Therefore, we will be concentrating our investigation on cardinal numbers and ordinal numbers.

The activity of counting involves a process called "mapping" by mathematicians. When mapping, we assign a separate object to each of the elements we are counting. In the common experience of using counting words, we assign to each item in the set being counted a specific counting word. The words must be used in the correct order, so that the word assigned to the last object becomes the cardinal number for the set. Hence, to count the fingers on my left hand I assign the following words to the fingers: one, two, three, four, and five. Five now is the cardinal number of the fingers of my left hand and tells me the manyness of that set.

Even though we generally use numbers when counting, this is not necessary. Some primitive societies have been known to use a kind of stick counting, where physical objects are mapped to the set of objects being counted. In such counting the counter does not end with a convenient word for his number of objects, but rather with a collection of objects that represents the manyness of the set. For example, a Wedda man from the island of Sri Lanka may count a stack of coconuts by assigning one clam shell to each coconut. When he is finished, he can't say how many coconuts he has for he has no word in his language to designate this number, but he can point to the pile of shells and say, "that many." If someone were to steal a coconut, he could perform the mapping of shells to coconuts again, and discover he had one extra clam shell. Holding the shell with no coconut to map it to, he would realize at once that one coconut was gone.

## ORIGINS OF COUNTING

When we count to find the number of elements in a set, we are using natural numbers as cardinal numbers. Therefore, the act of counting includes the desire to know the manyness of a set of objects. It is this desire that distinguishes true counting from simply speaking names in a designated order, such as in a song or poem. The desire to keep track of the number of elements in a set may have been the first motivation for human beings to count. How old is counting, and do other species count? One may be tempted to say that counting can be no older than language, since we typically use number words to count, but we have already seen that a more primitive form of stick counting can occur without language.

At the present time, no evidence exists that animals count. Elaborate experiments have been performed on birds and primates to test their counting talents.[2] The best they can do is to distinguish between the quantity of small groups of objects. They cannot map objects onto sets of elements to determine the manyness of the sets. The story is still out for dolphins and whales, both of which have brains large enough to carry out the mapping operation. While the human brain is approximately three pounds (1500 cubic centimeters in volume), a blue whale can have a brain as large as 15 pounds. We understand too little about such creatures to know if they can or do count.

The exact origin of counting is lost in our prehistory. Yet, it must be very old. The oldest direct evidence is a baboon's thigh bone discovered in the Lebembo Mountains of Africa that has been marked with 29 notches. The bone is 35,000 years old. A wolf bone found in Czechoslovakia has 55 notches and is 30,000 years old. While these bones are the products of *Homo sapiens sapiens*, or modern humans, their creation occurred long before farming (8000 B.C.), pottery (6500 B.C.), or wheeled vehicles (2700 B.C.). Humans living between 35,000 and 30,000 years ago were hunter–gatherers, with a Stone Age technology.

It's fun to speculate if counting could be even older than these 33,000–35,000-year-old animal bones. The humanoids preceding *Homo sapiens sapiens* were the Neanderthals who lived in Europe and the Middle East from approximately 130,000 years ago to around 35,000 years ago. No direct evidence exists from their living sites that these individuals counted, yet their average brain size was approximately the same as modern humans: 1500 cm³. They had the intelligence to use fire, make clothing and tools, create art, and bury their dead. It is not inconceivable that they also counted.

An even earlier candidate for counting is the next ancestor we encounter, moving back into our origins. *Homo erectus* was a hominoid living from approximately 1.5 million years ago to around 300,000 years ago.³ Early *Homo erectus* had a brain approximately 60 percent the size of our brains (900 cm³ compared to 1500 cm³), yet fully twice the size of a chimpanzee brain (450 cm³). Later *Homo erectus* brain reached a size of 1100 cm³ or almost 75 percent the size of our brain. Could the *Homo erectus* brain, midway between chimp and human, handle the counting operation? Certainly, the *Homo erectus* were clever. They mastered the use of fire, migrated out of Africa into Europe and Asia, made sophisticated stone tools, and constructed shelters. Such activities suggest the ability to conceive of a desired goal, and then carry out a set of specific activities to achieve that goal. This operation is precisely the kind of mental process needed for counting. We first form the desire to know the manyness of a collection of objects. Then, using either stick or language counting, we map objects or words against our collection. Therefore, the evidence surrounding *Homo erectus* suggests they may have had the mental capacity to count. But did they? Just having the ability to count does not guarantee that counting takes place. Several Stone Age tribes have been studied by anthropologists where the individuals do not count beyond two or three, classifying all larger collections as simply "many."

It is reasonable to assume that before counting began, there was a need for it. Hence, if *Homo erectus* counted, we can assume that counting increased their chances of survival in some manner. Just how could counting improve the survival of our ancient ancestors, be they *Homo erectus* or *Homo sapiens*? Certainly, within a cooperative hunting group, knowledge of how to divide the kill was important. If the game can be divided by some fair standard, the cooperation within the group is maintained. Hunting groups frequently traveled over long distances. If at least one of the group could divide the future into distinct days and plan for provisions, then the success of the hunt increased.

We can imagine the same kinds of demands for the domestic chores. How many days before the food on hand spoils? What proportions of different foods should be used in meal preparation? For at least one individual within the hunter–gatherer group to know how to count must have improved the group's success.

Without good physical evidence we cannot say the *Homo erectus* counted. Yet, we may get lucky at some future time when an anthropologist, struggling to uncover an ancient *Homo erectus* site in a desolate, wind-swept plain, stumbles across direct physical evidence of counting. This could be in the form of notched bones, collections of counting sticks or pebbles, or even notches on stone implements. One is tempted to go out into the backyard and toe over a few rocks to see what is underneath!

## TAKING THE NEXT STEP

Regardless of what we credit to old *Homo erectus*, we can say with confidence that modern humans were counting by 35,000 years ago. At one time anthropologists believed this period was the beginning of the *Homo sapiens* line, however new dating methods suggest that modern humans may have been around for 60,000 or even 90,000 years. If we were counting 35,000 years ago, we could have easily counted as long ago as 60,000 or 90,000 years ago.

What was the next great step in the evolution of numbers and counting? Farming began 10,000 years ago in the Fertile Crescent of the Middle East—that area from Jericho, north to southern Turkey, and then south down the Tigris–Euphrates valley to southern Iraq. The first farmers cultivated barley from which they could make bread and beer: the two basic units of trade in early Western Asia. Farming allowed a denser population to survive in a fixed area than did the lifestyle of hunting–gathering. With this settled lifestyle, a denser population grew, causing villages and towns to spring up.

Farming, and the associated towns and villages, presented new computing challenges for our early ancestors. Plots of land had to be measured off; grain had to be measured and stored; and workers had to be divided into work groups and eventually paid in beer and bread. Simple counting no longer sufficed to carry out all these chores. The pressure to account for the wealth which was accumulating in the towns and in possession of the kings and princes required new directions in arithmetic. The first response to this pressure was the use of clay counting tokens (Figure 2).

For many years archaeologists have been finding small clay objects of various shapes in the ancient sites of Western Asia. In the beginning, they supposed the tokens might be game pieces or fertility objects. But there were so many of them, and they were of such varied shapes. Could they all really be explained as toys or objects of worship? Finally, in the 1970s, Denise Schmandt-Besserat discovered their true use.[4] They were counting tokens, used to account for the various commodities accumulated by the early farmers. These tokens could be used much like pebbles, shells, and sticks mentioned before as aids in nonverbal counting. Yet, these tokens, as the products of production, could be manufactured in different shapes—shapes which took on specific meaning, and the shapes and styles could be standardized. This facilitated the evolution of a standard numerical system.

Sometime after the introduction of clay tokens, the early Sumerians, living in Southern Iraq, discovered that the tokens could be pressed into moist clay tablets before the tablets were baked. This provided the Sumerians with permanent records of objects counted. The next step was to draw the object being counted onto the clay along with the impressions of the tokens. Hence, five jars of oil could be recorded as five impressions from the "one" token, and a drawing of a jar of oil. This accounting system evolved into writing around 3100 B.C. in the major Sumerian cities of Southern Iraq.

The evidence we have of numbers from prehistoric times is, so far, limited to evidence for whole numbers. However, the early farmers of the Fertile Crescent could not carry out all their computing tasks if they had been limited to only positive integers. They needed fractions, and the numbering system of the early Sumerians

FIGURE 2.  Counting tokens from Western Asia.

demonstrates that they possessed a good grasp of division and fractions. Later, both the Babylonians, who conquered the Sumerians, and the Egyptians used fractions. The Babylonians used a numbering system based on both 10 and 60, and could easily record very large whole numbers and very small fractions. The Egyptian concept of fractions was somewhat more limiting since they only used unit fractions, i.e., fractions where the numerator was restricted to the number 1. Hence, all fractions had to be written as the sum of fractions such as 1/2, 1/3, 1/4, etc. This was an awkward and tedious requirement.

Soon after farming began in Western Asia, other civilizations began developing their own numbering systems and mathematics. Both China and Mesoamerica independently developed numbering systems. The Maya of the Yucatan Peninsula developed such an accurate mathematics to accompany their astronomy that they could predict the orbit of Venus to within a few hours over a period of 500 years.[5]

## FORMALLY DEFINING NUMBERS

We have talked about the natural number sequence, and we have identified it as a sequence that begins with the number 1 and then progresses through the natural numbers in the following manner: 1, 2, 3, 4, 5, . . . . This is the very sequence that we take great care to teach to our children, for it is the very basis of our arithmetic. We cannot show the last number in the sequence, because there exists no last number: the sequence "goes on forever." How do we know it really goes on "forever?" We can't see all the numbers written out, and how can anything exist that we cannot see?

Numbers are, of course, ideas, and we don't see ideas; we think them. But this is no help, because, if there exists an infinity of numbers, we cannot think of them all. This is the very problem that perplexed the early Greeks, and continues to perplex people today. Many people simply cannot accept the notion of an infinity of anything.

So far, I have offered no formally defined system for the natural numbers, but relied on our intuitive feeling for what they are and how they progress. Now it is time to consider such a formal system, called an axiomatic system. One of the first to offer such a system for natural numbers was the Italian mathematician Giuseppe Peano (1858–1932). That it took mathematicians so long to formalize the concept of numbers is not because it is a difficult job, but because numbers are so ingrained within us that mathematicians didn't feel the need for such a system for many millennia. That is, the nature of the counting numbers was "obvious."

An axiomatic system consists of a finite number of statements which when taken together, define the system. The statements contain words which are left undefined and simply accepted as primitive terms. Peano's axioms are simple, yet elegant.

Axiom 1: 1 is a number.

Axiom 2: The successor of any number is a number.

Axiom 3: No two numbers have the same successor.

Axiom 4: 1 is not the successor of any number.

Axiom 5: If 1 has a certain property, and the successor of every number has the same property, then every number has that property.[6]

What beautiful statements for our counting numbers. The axioms mention "numbers," "successor," and the number "1." These are our primitive terms, and we don't offer any formal definition of them. Yet, they are meant to have the normal meaning and refer to the natural number sequence. The first axiom simply states that 1 is a number. The fourth axiom says that 1 is not the successor to any other number. This makes 1 the first number in our sequence, just as it should be. The second axiom says that the successor of any number is also a number. Hence, each succeeding term in the sequence is another number. This ensures that the sequence does not break down at some point and start producing things other than numbers, such as lines, or angles, or watermelons. Axiom 3 guarantees that no two different numbers have the same successor. Why should this be important? Suppose we used the following sequence for our numbers: 1, 2, 3, 4, 2, 5, 6, 7, . . . . Notice that both the numbers 1 and 4 have the same successor number, i.e., the number 2. If we were to use such a number sequence for counting, then when we completed a count of a set with the number 2, we would not be sure of the actual number of objects (elements) in our set. That is, does the set we are counting contain two or five elements? Hence, we see that each and every number must be unique. Axiom 3 gives us that guarantee.

The first four axioms specify those very properties of the counting sequence that we learn from childhood: 1 is the first number, each number has a successor which is also a number, and no number is the successor of two other numbers, i.e., the numbers are all unique.

The fifth axiom requires a little more thought. These five axioms were intended to be the logical foundation for all of arithmetic. For the axioms to have such power, it was necessary for Peano to include this last axiom, known as the axiom of induction. This axiom allows us to prove things about all numbers, even though there are an infinite number of them. Let's say I can prove something about the number 1. Suppose I can also prove the same thing about the number 2. Let's pretend I can continue and prove the property for the number 3. Is this property true for all numbers? I continue and prove it for 4, 5, and 6. But no matter how many times I extend my proof, I've still only proved the property for a finite number of numbers. Axiom 5 allows me to extend my proof to all numbers so long as I can prove that successors always have the property in question. Thinking about this axiom, we see that it agrees with common sense (which is, of course, no guarantee that it is correct). If we can prove something for 1 and for all successor numbers, we prove it for all numbers since all numbers are either successors, or the number 1. We will use this axiom in a proof when we consider some of the early Greek discoveries about numbers.

Remarkably, these five axioms are sufficient to define all the operations of arithmetic, and to prove many of the wonderful theorems dealing with numbers. We define the number 2 as the successor of 1, and we show it symbolically as $2 = 1 + 1$. Three is simply the successor of 2, or $3 = 2 + 1$. The operation of addition can be defined in terms of successors, for each number's successor is just that number with 1 added to it. For example, if we have 2 + 3 we can define the result as a combination of successors.

| | |
|---|---|
| Step 1: write 3 as the successor of 2: | $2 + 3 = 2 + (2 + 1)$ |
| Step 2: the second 2 is the successor of 1: | $2 + (1 + 1 + 1)$ |
| Step 3: rearrange the 1s: | $(2 + 1) + (1 + 1)$ |
| Step 4: write the successor of 2 as 3: | $3 + (1 + 1)$ |
| Step 5: rearrange the 1s again: | $(3 + 1) + 1$ |
| Step 6: write the successor of 3 as 4: | $4 + 1$ |
| Step 7: write the successor of 4 as 5: | $5$ |

Admittedly, the process is slow and tiresome. Yet, from this basis, we move on to define subtraction as the inverse operation of addition, then multiplication as successive addition, and finally division as the inverse operation of multiplication. All the wonderful characteristics of arithmetic come tumbling out of Peano's five axioms.

Even though the key terms within the axiomatic system are undefined, the axioms specify that they stand in a certain relation to each other. It is this wonderful relation that gives us the natural numbers. We no longer are required to laboriously add 1 to each number in the sequence to generate the next number, and therefore limit the numbers we can talk about to those we have actually defined and written down. The axioms specify the desired relations, and all our infinity of natural numbers come popping into existence at once. Because of the way we learn numbers as little children—that is, always beginning with 1 and counting up to 10 or more—we have the impression that the first numbers, those below 10, have some higher order of existence than the numbers which follow. The ancient Greeks recognized this intuitive feeling by stating that the decade, or 10, was a sacred number. One such Greek was Philolaus of Tarentum who said:

> One must study the activities and the essence of Number in accordance with the power existing in the Decad (*Ten-ness*); for it (*the Decad*) is great, complete, all-achieving, and the origin of divine and human life and its Leader;. . .[7]

Yet, there is no valid reason to hold that any one number has more of a claim to existence than any other. To say that 3, 5, or 9 have more existence than say, 17, 73, or even 3982, places an unnecessary condition on the notion of existence. In whatever way numbers do exist, they share in existence equally. This is not to say that all numbers share all the same qualities. Some numbers do, indeed, have unique and fascinating characteristics, and it is the discovery of these characteristics that makes the study of mathematics so intriguing. Our entire adventure into mathematics will be to uncover the characteristics of individual numbers and the relationships between collections of numbers.

## EARLY GREEK ACHIEVEMENTS

While older, more established societies used numbers and mathematics before the Greeks, the Greeks get credit for much early work in mathematics because they took the time to record their discoveries. In all likelihood, some of these discoveries were made by earlier civilizations, but their mathematicians either failed to record them (possibly keeping their knowledge secret) or their writings have been lost to us. The Greeks, however, left a great and rich tradition of literature for us covering science, philosophy, drama, and mathematics. Within mathematics, they were able to formulate questions regarding numbers, some of which have not been answered to this day.

The first human to be identified as having made a contribution to mathematics was Thales of Miletus (634–548 B.C.). Miletus was a Greek city located on the west coast of Asia Minor with trading connections to both the more ancient civilizations of Babylonia and Egypt. Thales established a school in Miletus where he taught mathematics and philosophy, and was considered by later Greeks to be the first of the seven wise men of Greece. While he is given credit for a number of specific discoveries in geometry, it was his use of the deductive method which was his greatest achievement. Beginning with known truths, he proceeded to deduce new truths. This method so impressed Thales' countrymen that they adopted deductive reasoning as the hallmark of Greek thinking. Hence, the deductive method came to be the leading characteristic of mathematics.

However, Thales is not the central character of our quest. The man who would come to outshine even Thales was Pythagoras from the Island of Samos (580–500 B.C.). Pythagoras (Figure 3) is reported to have attended Thales' school in Miletus and may have even received instruction from Thales himself. He, like Thales, is reported to have journeyed to both Egypt and Babylon in his early years to receive instruction from the priests of these much more ancient civilizations. In his later years Pythagoras, too, established a school which was located in Croton in Southern Italy.

FIGURE 3. Pythagoras, ca. 580–500 B.C.

The ideas of Pythagoras went beyond those of Thales. Whereas Thales considered his objects of mathematics—numbers, lines, and angles—as objects of thought and not physical entities, Pythagoras extended the idea of number by claiming that not only were numbers objects of thought, they were also the building blocks of all reality. Where could he have gotten such an idea? One of his discoveries was the relationship between the length of a taut string and the sound the string produces when it is plucked. He noticed that if a string is shortened to 1/2 its original length, then the tone produced is one octave higher than the original tone. Hence, lengths of string that were in a ratio of 1:2 produced tones that were in harmony. It was this discovery of the musical intervals, including the octave with ratio 2:1, the fifth with ratio 3:2, and the fourth with ratio 4:3, that inspired the Pythagoreans to hold the number 10 as sacred, since the four numbers used in these ratios, i.e., 1, 2, 3, and 4, add up to 10.[8]

This discovery greatly influenced Pythagoras. Evidently he didn't stop to consider the role that the strings, themselves, played in producing the sound, e.g., the vibration of taut strings and the design of our ears. He made the assumption that the harmonic tones were due only to the proper ratio of the strings' lengths. Hence, number ratio was what was really important. From this he concluded that all material objects in the universe owed their natures to the nature of number. Numbers, in fact, are the atoms of the universe, combining to form everything else.

> Evidently, then, these thinkers [Pythagoreans] also consider that
> number is the principle both as matter for things and as forming
> both their modifications and their permanent states. . .[9]

This idea of number being the generator and prime cause of everything else became a hallmark of the Pythagorean order. Even though Pythagoras established a religious–philosophical order where discoveries were keep secret, after his death his disciples spread throughout the Greek world and, by establishing their own schools, insured the spread of his ideas. The father of idealism, Plato, was greatly influenced by the Pythagoreans of his day, and expanded on the Pythagorean belief that number was the ultimate cause to conclude that a hierarchy of ideas, or Forms, were the cause of the universe. At the top of Plato's hierarchy was the form for "The One" or "The Good," which became Plato's prime generating agent of reality.

The practice of keeping mathematical discoveries secret within a caste of loyal priests and scribes was not unique to the Pythagoreans. In the ancient world, the ability to use mathematics to predict celestial events and the four seasons was a critical kind of information for any king or emperor. Farmers had to be informed by their king when to plant their crops. The flooding of rivers and eclipses of the sun and moon had to be predicted. A king who failed to warn his subjects of oncoming floods or eclipses could quickly lose popular support. In 17th century B.C., Ahmes, an Egyptian scribe, wrote the Rhind Papyrus, one of the oldest documents on mathematics, and spoke of his writing as "knowledge of existing things all, mysteries . . . secrets all."[10] When knowledge meant power, it had to be jealously guarded.

Even today some believe in the ancient Pythagorean notion that numbers control and influence all things. We see this in both numerology, the occult practice of decoding names and birthdays into numbers, and gematria, decoding religious scriptures into numbers. But just what were these discoveries of the Pythagoreans and the other ancient Greeks that led them to the belief in the power of numbers? Civilizations had been using basic arithmetic for centuries before the Greeks, for the Greek merchants possessed a good understanding of the operations of addition, subtraction, multiplication, and division. The ancient Greek thinkers actually considered the four operations of arithmetic (which they called *logistic*) to be inferior to their own study of number theory (which they called *arithmetic*). Number theory is the study of the characteristics of numbers rather than how to manipulate them in computations. Today, number theory has evolved into a deep and profound field within mathematics, and as such, is the primary focus of our investigation.

The early discoveries of number theory appear to be quite simple, but this is deceptive, for we'll see that a number of simple questions asked by the leading Greek mathematicians of yesteryear have still not been answered.

## EARLY DISCOVERIES

The Greeks were interested in the counting numbers. They did not know of negative numbers and considered fractions to be only ratios between natural numbers rather than numbers in their own right. The oldest characteristic discovered about natural numbers is that they are either even or odd. An even number, of course, is any number that can be evenly divided by 2. An odd number is any number that cannot be so divided. Beginning with 1, every other number is an odd number. Beginning with 2, every other number is even. This distinction between even and odd plays a fundamental role within mathematics.

An early custom of the Greeks, called pebble notation, was to use sets of pebbles to represent numbers. Different numbers of pebbles could be arranged on the ground in different shapes. For example, the pebbles representing the numbers 3, 6, and 10 can be laid out in the form of triangles (Figure 4). Hence, these numbers became "triangular numbers." The Greeks also noted that if they calculated consecutive sums of the natural numbers as they appear in the number sequence, they always got triangular numbers. Thus, if we add 1 and 2 we get the triangular number 3. This process can be extended without end.

1 + 2 = 3

1 + 2 + 3 = 6

1 + 2 + 3 + 4 = 10

1 + 2 + 3 + 4 + 5 = 15

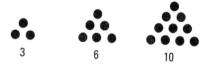

FIGURE 4.  Pebble arrangements for the triangular numbers 3, 6, and 10.

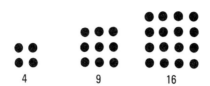

FIGURE 5.   Pebble arrangements for the square numbers 4, 9, and 16.

This law of forming triangular numbers may go all the way back to Pythagoras himself.[11]

Are there other special kinds of numbers besides triangular? Yes! Some numbers in pebble notation can be laid out in the form of a square. In Figure 5 we have the pebble representations for 4, 9, and 16. Such numbers were designated as "square" numbers. The Greeks discovered that if they added consecutive odd numbers they always got square numbers. We can write such numbers in several convenient forms.

$1 = 1 \cdot 1 = 1^2$

$1 + 3 = 4 = 2 \cdot 2 = 2^2$

$1 + 3 + 5 = 9 = 3 \cdot 3 = 3^2$

$1 + 3 + 5 + 7 = 16 = 4 \cdot 4 = 4^2$

When we see the pebble formation of square numbers we understand how the discoverer (possibly Pythagoras) realized this law. Figure 6 shows a sequence of square numbers beginning with 1 and progressing to 36. Each higher square number is formed by adding an L-shaped set of pebbles to the previous number. Hence, 4 is constructed by adding the L-shaped set of three pebbles to one pebble. The next square number, 9, is formed by adding the L-shaped set of five pebbles to the square number 4. To get each succeeding square number we just add an L-shaped set containing the next odd number of pebbles. This L-shape was called the *gnomon* by the Greeks, and originally referred to an instrument imported to Greece from Babylon and used for measuring time. To get the value of a square number, $n^2$, we add all odd numbers up

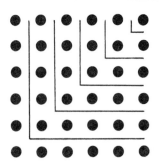

FIGURE 6.   Building a square number from a smaller square number by adding an L-shaped (*gnomon*) set of pebbles.

to $2n - 1$. Hence, to calculate the square of 5 or $5^2$ we add all odd numbers up to $2 \cdot 5 - 1$ or 9. Thus, we have: $5^2 = 1 + 3 + 5 + 7 + 9 = 25$. To form the next square number beyond 25 all we do is add $2n + 1$. To get $6^2$ we add $2 \cdot 5 + 1$, or 11, to 25 resulting in the new square number 36.

Of course, by computing the first few sums of odd numbers and seeing that these sums are perfect squares, we do not prove that it is always the case. Could there be some large value of $n$ such that when we add all the odd numbers together up to $2n - 1$ we get a number that is not a perfect square? To prove this situation will never happen we can use Peano's fifth axiom. We have already demonstrated that the sum of odd numbers is a perfect square when $n = 1$. Now we will assume that it is also true for some $n$ which we don't specify. We can do this because we know it's true for $n = 1$. Hence, we say that for some $n$:

$$1 + 3 + 5 + \ldots + (2n - 1) = n^2$$

On the left of this equation we have the sum of all odd numbers up to $2n - 1$ and on the right we have a perfect square, namely $n^2$. Can we now show that this same characteristic holds for the successor of $n$, or $n + 1$? To do this we will take the above equation, which we assume to be true, and add $2n + 1$ to both sides. Adding the same value to both sides of an equation does not, of course, invalidate the equality.

$$1 + 3 + 5 + \ldots + (2n - 1) + (2n + 1) = n^2 + (2n + 1)$$

On the left side of the equal sign we can rewrite $2n + 1$ as $[2(n+1) - 1]$ while on the right we can factor the terms into $(n + 1)^2$. This yields:

$$1 + 3 + 5 + \ldots + (2n - 1) + [2(n + 1) - 1] = (n + 1)^2$$

What we have demonstrated here is that if the characteristic is true for $n$, then it is also true for $n + 1$. Hence, from Peano's axiom the characteristic must hold for all numbers $n$. Therefore, we've used the axiom of induction to prove that the sum of all odd numbers up to $2n - 1$ will always equal $n^2$.

This discovery about odd numbers and perfect squares must have astounded the Greeks, and suggested to them that numbers truly did contain some great magic, enabling them to generate triangular and square numbers without end. This ability of numbers from the number sequence to "generate" other numbers of a certain kind could lead to the belief that numbers generate much more, an idea which fit nicely with the Pythagorean idea that numbers were the prime cause of the universe. As we proceed, we'll discover that the number sequence truly is a powerful tool for generating additional mathematical concepts.

In mathematics it is always interesting to try to expand on an original idea. Using the odd numbers in the number sequence again, we can form the cube numbers. To get $n^3$ we add successive sets of $n$ odd numbers.

$1^3 = 1$

$2^3 = 8 = 3 + 5$

$3^3 = 27 = 7 + 9 + 11$

$4^3 = 64 = 13 + 15 + 17 + 19$

$5^3 = 125 = 21 + 23 + 25 + 27 + 29$

Notice that to get the cube of 1, we started with just the first odd number, and to get the cube of 2 we added the next two odd numbers. The cube of 3 is generated by adding the next three odd numbers, etc.

The Greeks discovered even more kinds of numbers. Not all numbers can be put into the form of a triangle or a square. If we

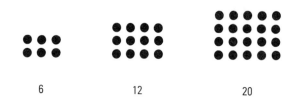

6                    12                    20

FIGURE 7.   Pebble arrangement for the oblong numbers 6, 12, and 20.

add consecutive even numbers beginning with 2, we get numbers that can be formed into a rectangle where the length of the sides differ by one (Figure 7). Such numbers were called oblong. Notice that the number 6 can be put into the form of both a triangle and an oblong. Hence, some numbers possess numerous characteristics. The Pythagoreans also discovered that every oblong number was the sum of two triangular numbers (Figure 8).

We now arrive at a very special kind of number. While playing with their pebble notation the Greeks discovered that certain numbers cannot be formed into a rectangle or a square. No matter how we try, there is always one pebble left over. They called these numbers linear, but we know them today as prime numbers. Notice that with all oblong and square numbers, the numbers can be factored into two smaller numbers. Hence, 12 becomes 3 times 4 or 3·4. Whenever we factor a number into two smaller numbers we can build a rectangle with the smaller numbers representing the number of pebbles on each side. Therefore, 12 can be arranged as a rectangle with three pebbles on the vertical side and four on the horizontal side. We can think of this as 4 representing the number of pebble columns and 3 representing the number of pebble rows.

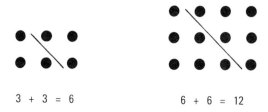

3 + 3 = 6                    6 + 6 = 12

FIGURE 8.   Pebble arrangement demonstrating that each oblong number is the sum of two triangular numbers.

However, our linear or prime numbers cannot be arranged into rectangles, and they cannot be factored into smaller numbers. The definition of a prime number is stated in terms of those whole numbers which will evenly divide it.

> *Prime number*: A number, excluding 1, that can only be evenly divided by the number 1 and itself.

Two is the first prime number. The number 1 technically satisfies our definition, yet we exclude it for a reason which will soon be obvious. Two, the first prime number, is also the only even prime number, for any other even number can be evenly divided by 2. Hence, 2 is a very special number. The next prime is 3, for no matter how we try we can't make a rectangle out of 3 (although 3 is a triangular number). Five is the next prime followed by 7, 11, 13, 17, and 19.

All the natural numbers in our number sequence divide themselves into prime numbers, and nonprime numbers, called composite numbers. Composite numbers, such as 10, 15, 18, and 45, can always be factored into the product of two smaller whole numbers. This division between prime and composite numbers turns out to be one of the cornerstones of mathematics, and is a characteristic which is used in mathematical proofs over and over. As we will see, some of our most profound questions in mathematics involve prime numbers.

To understand the importance of the characteristic of being either prime or composite, we must look at how composite numbers factor. If we factor a composite number into two smaller numbers, then we can ask whether these two numbers are, themselves, prime or composite. For example, 6 factors into 2·3. Both 2 and 3 are prime numbers. Eighteen factors into 2·9. Here, the 2 is prime but the 9 is not. In turn, 9 factors into 3·3, and 3 is a prime number. Hence, 18 factors into the following prime numbers: 18 = 2·3·3. Beginning with any composite number, no matter how large, we can, because it is composite, factor it into two smaller numbers. We then ask whether each of the smaller factors are prime or composite. If either one is composite we factor it again, and then ask the question once more. Sooner or later, we'll factor the original composite number into just prime numbers. Hence, every composite number can be written as the product of prime numbers. Since prime numbers are numbers which are already prime, we can say that all whole numbers from the natural number sequence can be represented as prime numbers or the product of prime numbers.

This, in itself, is interesting, yet it leads to an even more fascinating law. It turns out that when a composite number is factored into primes, those particular primes are unique to that number. For example, we can factor the number 30 into 2·3·5. No other set of prime numbers, when multiplied together, will yield 30. We can certainly rearrange the numbers such as 3·2·5 or 5·2·3, yet we must always use these three specific primes to get 30. This leads to one of the great building blocks of mathematics.

> *Fundamental Theorem of Arithmetic*: Every whole number greater than 1 can be expressed as a product of prime numbers in one and only one way.

This is why we excluded the number 1 from being a prime. If we let 1 be a prime number then we do not have our unique factorization and Fundamental Theorem of Arithmetic. For example, 6 could factor into both 2·3 and 1·2·3. Hence, the factorization of 6 would not be unique. This characteristic of unique factorization represented by the fundamental theorem is heavily used throughout mathematics.

## PERFECT AND FRIENDLY NUMBERS

Now that we've introduced the notion of factoring, we can demonstrate additional discoveries of the Greeks. Every composite number can be evenly divided by each of its primes. It can also be evenly divided by any combination of its primes as long as we restrict ourselves to the number of primes found in the original factorization. For example, 12 factors into 2·2·3, where we have two 2s and one 3. Twelve, therefore, is not only divisible by 2 and 3, but by 2·3 (6) and by 2·2 (4). Hence, the numbers evenly dividing 12 are 1, 2, 3, 4, 6, and 12. Here we have included 1 as an even divisor of 12 even though it is not a prime.

Realizing this, the Greeks asked: What kind of number do you get when you add the divisors of a composite number, excluding the number itself? For example, we discovered that 12 can be evenly divided by 1, 2, 3, 4, and 6. We add these together to get 16. The number 16 is larger than 12, and therefore we call 12 abundant. It is abundant because its divisors add to more than the number itself. If we look at the divisors of 10 we get 1, 2, and 5. These add to 8 which is less than 10. Because 8 is less than 10, we call 10 deficient. Are there numbers that are neither abundant nor deficient, i.e., numbers whose divisors add to the number itself? Consider 6: Its divisors are 1, 2, and 3, and these three numbers add to 6. The Greeks called 6 a perfect number. We see that 6 is a very unique number because it is not only the first composite number which is not a square but is also the first perfect number.

Are there more perfect numbers? The next is 28 which is the sum of 1 + 2 + 4 + 7 + 14. To find the next perfect number we must jump all the way up to 496 = 1 + 2 + 4 + 8 + 16 + 31 + 62 + 124 + 248. For the next perfect number we must sift through all the numbers up to 8128. Euclid (Figure 9), living at the end of the third century B.C., was the first to mention these numbers, and Nicomachus of the first century A.D. knew of the first four: 6, 28, 496, and 8128. The next perfect numbers are: 33,550,336 and 8,589,869,056. By inspection we see that the perfect numbers are growing very large. In fact the ninth perfect number contains 37 digits.

All the perfect numbers discovered to date are even. These follow a definite form, and to understand this form we introduce the notation of exponents. When we square a number $A$, we can write it as $A \cdot A$ ($A$ multiplied by itself). A second method of showing this product is to write $A$ with a small 2 as a superscript: $A \cdot A = A^2$. We can now generalize this notation. If we multiply a number $A$ by itself $n$ times we represent it as $A^n$. This convenient notation makes it easy to write certain large numbers.

All even perfect numbers have the form $2^{n-1}(2^n - 1)$, when the number $(2^n - 1)$ is itself a prime number and $n$ is also prime. Let's try this formula when $n = 2$:

$$2^{2-1}(2^2 - 1) = 2^1(4 - 1) = 2 \cdot 3 = 6$$

FIGURE 9.   Euclid (fl. 300 B.C.)

Hence, because $2^2 - 1$ was a prime number (3) and 2 is prime, the formula produced 6, our first perfect number. Let's now try $n = 3$:

$$2^{3-1}(2^3 - 1) = 2^2(8 - 1) = 4 \cdot 7 = 28$$

We get our second perfect number. Unfortunately, this formula does not work in those cases when $n$ turns out not to be prime. This occurs when we try $n = 4$, a composite number.

$$2^{4-1}(2^4 - 1) = 2^3(16 - 1) = 8 \cdot 15 = 120$$

But 120 is not perfect. This realization led the Greeks to ask: How many perfect numbers exist, and are any of them odd? No one knows. The problem is, we don't know if an infinite number of primes of the form $(2^n - 1)$ exist where $n$ is also prime. Such primes are called Mersenne primes after Marin Mersenne, a Minimite friar (1588–1648). The largest known Mersenne prime is $2^{756839} - 1$ and was discovered in 1992 with the use of a Cray-2 computer.[12] This is a number formed by multiplying 2 by itself 756,839 times and then subtracting 1. Using this largest known Mersenne number we can form the largest known perfect number: $2^{756838}(2^{756839} - 1)$. This is, indeed, a very large number containing 455,663 digits if written out in normal form. That's enough digits to take up almost 180 pages if written in a book this size.

We don't know if an infinite number of Mersenne primes exist or whether an infinite number of perfect numbers exist. Nor do we know if any odd perfect numbers exist, although it has been verified by computers that none exist that are smaller than $10^{300}$ (i.e., 1 followed by 300 zeros).[13] Part of the Greek genius was their ability to pose such deep questions that after 2000 years we are still struggling for answers.

Another property of numbers, probably discovered by Pythagoras, was friendliness. Two numbers are friendly to each other if the sum of their divisors equal each other. An example would be 284 and 220, for we have:

$$284 = 1 + 2 + 4 + 5 + 10 + 11 + 20 + 22 + 44 + 55 + 110$$

and

$$220 = 1 + 2 + 4 + 71 + 142$$

With a handy calculator it is easy to verify that the numbers adding to 284 are, in fact, all the divisors of 220, while those adding to 220 are the divisors of 284.

There are many more questions about whole numbers which were first suggested by the ancient Greeks, and some of these will be considered later.

## BIG NUMBERS

We have seen that the natural number sequence, as defined, has no largest number, but continues indefinitely to ever larger numbers. This leads to speculation regarding the size of numbers. Most numbers we use in normal daily activity reach into the hundreds and even thousands. These are the numbers we use when paying bills, buying groceries, and balancing our checkbooks. Through newspapers and television we are exposed to even larger numbers. We hear that different government programs cost millions or even billions of dollars, and that our national debt is now over five trillion dollars. These are, indeed, large numbers, and compared to the hundreds or thousands we are used to dealing with, they are difficult to fully comprehend. A million is a thousand thousand or 1,000,000. A billion is a thousand million, or 1,000,000,000.

To get some perspective on these numbers it helps to relate them to something in the real world. If the average human heart beats 72 times a minute, then it will beat 37,843,200 times in one year. Hence, we could count to a million by simply counting our pulse beats for about 9 1/2 days. Possible, but certainly too boring to do. How long would it take to count to a billion? It would take 26 1/2 years! How many times will our hearts beat if our average age is 74 years? About 2,800,000,000 times.

The numbers are getting so large now that it will be convenient to introduce a new notation, which is really an extension of our exponential notation. If $100 = 10 \cdot 10$ then $100 = 10^2$. In like manner $1000 = 10^3$. Extending this idea we can write one million as $10^6$ and one billion as $10^9$. When we see $10^9$ we realize that the number is a 1 followed by nine zeros. If we want to represent the number of lifetime heartbeats or 2,800,000,000 we can rewrite it as $2.8 \times 10^9$. This product stands for the number 2.8 multiplied by a number formed from 1 followed by nine zeros. This method of representing numbers is known as scientific notation.

We can also represent small numbers with this notation. The decimal 0.1 (which is the same as 1/10) can be written as $10^{-1}$, and in a similar manner we can write 0.01 (1/100) as $10^{-2}$. Hence, the number $10^{-n}$ is a decimal with a 1 located $n$ places to the right of the decimal point. Therefore we can show very small numbers with a corresponding large $n$.

On a practical scale we have almost reached the limit of numbers used by the general population. Our national debt is approximately 5.5 trillion ($5.5 \times 10^{12}$), where one trillion is one thousand billion. How long would it take to count up to our national debt if we counted at the rate of 72 per minute? Approximately 145,000 years.

Larger numbers are primarily used by scientists. We have actually given names to numbers beyond one trillion, but there is so little use for them that they are generally unfamiliar to most of us.[14]

| | | |
|---|---|---|
| $10^2$ | = 100 | hundred |
| $10^3$ | = 1,000 | thousand |
| $10^6$ | = 1,000,000 | million |
| $10^9$ | = 1,000,000,000 | billion |
| $10^{12}$ | = 1,000,000,000,000 | trillion |
| $10^{15}$ | = | quadrillion |
| $10^{18}$ | = | quintillion |
| $10^{21}$ | = | sextillion |
| $10^{24}$ | = | septillion |
| $10^{27}$ | = | octillion |
| $10^{30}$ | = | nonillion |
| $10^{33}$ | = | decillion |
| $10^{36}$ | = | undecillion |
| $10^{39}$ | = | duodecillion |
| $10^{42}$ | = | tredecillion |
| $10^{45}$ | = | quattuordecillion |
| $10^{48}$ | = | quindecillion |
| $10^{51}$ | = | sexdecillion |
| $10^{54}$ | = | septendecillion |
| $10^{57}$ | = | octodecillion |
| $10^{60}$ | = | novemdecillion |
| $10^{63}$ | = | vigintillion |

Other than the observation that some of the names are amusing, their usefulness is questionable since we don't encounter numbers over a trillion frequently enough to have need of a name for them. Scientists prefer to use the scientific notation and dispense with the use of proper names. Yet, people have gone on to name even larger numbers. For example, the number $10^{100}$ (a 1 followed by 100 zeros) has been named a googol. Written out it becomes:

10,000,000,000,000,000,000,000,000,000,000,000,000,000,
000,000,000,000,000,000,000,000,000,000,000,000,000,000,
000,000,000,000

But the googol isn't the largest named number for we have $10^{googol}$ which is called a googolplex. That's a 1 followed by a googol of zeros. If we assume a normal 300-page book can contain approximately 800,000 digits, a googolplex would fill $1.25 \times 10^{94}$ volumes if written in normal form.

Are the googol and googolplex larger numbers than we really need? In the physical world we can compare the very large to the very small to see how big of a number we can generate. The radius of a hydrogen atom is approximately $1.74 \times 10^{-10}$ feet while the radius of the known universe is approximately $10^{26}$ feet. If we divide the radius of the universe by the radius of the hydrogen atom we discover the number of hydrogen atoms, laid out end to end, that would reach across the universe. The number is $5.75 \times 10^{35}$. This is not even close to a googol. Another large physical ratio is the weight of an electron at $2 \times 10^{-30}$ pounds compared to the mass of the universe at $4 \times 10^{52}$ pounds. Dividing the mass of the universe by the mass of the electron we get $2 \times 10^{82}$. Still not a googol.

To demonstrate the largest numbers used today, we must leave the examples from the physical universe and consider numbers encountered in mathematics. For many years the largest number considered to be useful to mathematics was Skewes' number, a number which we will later see comes from a deep question concerning primes:

$$Skewes'\ number \approx 10^{10^{10^{34}}}$$

Skewes' number is definitely larger than a googolplex. This is easy to see when we write both the googolplex and the Skewes' number in the same format:

$$googolplex = 10^{10^{10^2}}$$

However, we can do even better. A recent mathematics thesis[15] on the number of kinks in the core of an embedded tower gives an estimate of:

$$10^{10^{10^{10^{10^{10^7}}}}}$$

Now, that's a big number!

## HOW BIG CAN WE GO?

We can, of course, continue to define even bigger numbers, since there is no end to the numbers available to us. The numbers we have been dealing with in scientific notation are specific numbers, yet they are all powers of ten. This makes them easy and compact to write. However, if we were to pick a number at random between the $10^{100}$ and $10^{101}$ we would, in all probability get a number whose digits varied randomly. Hence, to record such a number we must write down all 100 digits. When we progress to the random numbers as large as the largest known prime, we encounter numbers which require hundreds of thousands of digits to write them out exactly. Such numbers would fill a normal book of 100 pages, and the task of recording and handling these numbers is daunting. If we were to consider a random number in the same range as Skewes' number, we would get a number which is physically too large to even attempt to write. If we record a number as large as Skewes' number in books, each book weighing two pounds, with only $4 \times 10^{52}$ pounds of mass in the universe, there just isn't enough material in the entire universe to do the job. We must also point out that only a finite (although very large) number of integers exist that are smaller than Skewes' number, while an infinite number exist that are larger. This means an infinite number of numbers exist which we cannot even write down. In fact, comparing the numbers we do deal with to all the infinity of numbers, the numbers we can handle are an infinitesimal fraction of all numbers (if, indeed, the term "infinitesimal fraction" has any meaning).

This leads to an interesting enigma. Suppose a way existed to randomly select a number from all the infinity of counting numbers. How could we describe such a number? We can suppose the number is big. Let's pretend that to write it down on a piece of paper, the paper would fill the entire universe. Even though the number, by our standards, is large, compared to all the numbers that exist that are larger, it is still exceedingly small. In fact, it is so small that the chances of choosing such a small number are almost nonexistent. A randomly chosen number must be much larger. But how large? This is where we begin to get into trouble, for no matter what number we come up with as an example, we immediately see there exists only a finite number of numbers that are smaller, and an infinity of numbers that are bigger. Hence, the probability of ever randomly choosing such a small numbers is, for practical purposes, nonexistent.

The problem may be in our notion of selecting just one number at random from an infinite set of numbers. The possibility exists that this cannot be done; that to choose a number by any means requires us to specify some information about that number. But let's put such reservations aside, and pretend we can make random selections. We still have the interesting question of whether we can make any statements about such an unbelievably large number. For example, we can say the probability of the number being an even number is about 50:50. We know this because every other number is even. What would the probability be of the number being a prime? Can we make any guess about how many factors it might have? Do other characteristics exist that might be useful to describe such a number?

As it turns out, we can make meaningful statements about such numbers, and we will consider this question again as we progress on our fascinating trip along the number sequence.

## CHAPTER TWO

# NUMBERS AND THE OCCULT

*The Martians seem to have calculated their descent*

*with amazing subtlety—their mathematical learning*

*is evidently far in excess of ours.*

H. G. WELLS

*THE WAR OF THE WORLDS*[1]

## POWER WITHIN NUMBERS

*C*entral to the use of numbers within magic and occult practices was the ancient belief that numbers, by themselves, had power to influence the corporeal world, a concept universally rejected by the modern scientific community. How did such a strange idea come about in the first place? If we take a moment to consider the conditions within the ancient world, we can come to appreciate how the motivation to believe in the power of numbers must have been very strong.

In early times science was young, and most natural phenomena could only be explained as the action of the gods. As people began to farm, they realized that the seasons changed, and these changes were reflected in the different locations of star groups in the sky. As a result, they saw the connection between seasons, the production of their food (their very survival), and the stars—astronomy. Since the stars moved and the seasons changed, it was easy to believe the stars somehow caused the changes here on earth—astrology. In fact, any clever person who could record past seasons and star positions could then actually predict the coming seasons. There-fore, numbers, used to record the dates of past events, became the connection between earthly matters and the stars. This suggested to the ancients that a causal connection existed between numbers and worldly happenings. Those who could use numbers to calcu-late forthcoming events, such as the next planting season, flooding on the Nile, or eclipses of the sun, gathered great power to them-selves. They became the wizards, mathematicians, and astrono-mers in the courts of kings. If knowing numbers could give one such power, some connection must exist between numbers and individuals who wield this power—hence the logical idea that numbers influence us as individuals. Unfortunately, just finding a logical connection leading to a new idea does not automatically make that idea true.

The Pythagoreans of the Greek world elevated the importance of numbers to new heights. Because of the supposed connection between such objects as stars, numbers, and their daily lives, the Pythagoreans evolved a metaphysics based on number which was also connected to other, nonnumerical attributes. Table 1 shows that the Pythagoreans separated attributes into two broad categories. In the beginning, they supposed, the principle of the *unlimited* mixed with the *one* and separated out numbers which then became the building blocks of the universe. Beginning with such a metaphys-ics, it was then possible to borrow individual attributes from either

**Table 1.**  Pythagorean Metaphysical
Attributes

| Limit | Unlimited |
|---|---|
| odd | even |
| one | plurality |
| right | left |
| male | female |
| resting | moving |
| straight | curved |
| light | darkness |
| good | bad |
| square | oblong |

the limit or the unlimited and apply them to their personal lives. The Pythagoreans were probably not the first to practice this type of pseudoscience by association, nor were they the last.

## NUMBERS AND MAGIC

While numbers have historically played a role in the general practice of magic, this role was somewhat limited. However, when we investigate the specific occult practices of numerology and gematria, we will see that numbers become the unifying theme.

In the overall belief system of magicians, certain mystic powers were associated with individual numbers, usually those numbers between 1 and 10. When these numbers were used within specific magic rituals, the mystic power transferred from the numbers to the user or to the object of the ritual. For example, the number 7 was frequently associated with both good luck and spirituality. Many examples of the number 7 used in ritual can be found in the Old Testament of the Bible. For example, as part of God's instructions to Moses for priests making a blood offering we find:

> And the priest shall dip his finger in the blood, and sprinkle of the blood seven times before the Lord, before the veil of the sanctuary.[2]

It is interesting to note that God took six days to create the world, and then rested on the seventh. When it was necessary for Joshua to take the walled city of Jericho, the Lord instructed him:

> And seven priests shall bear before the ark seven trumpets of rams' horns: and the seventh day ye shall compass the city seven times and the priests shall blow with the trumpets.[3]

Certainly, the number 7 must have held special power for the Hebrews, for the Lord did not say, "A bunch of priests shall walk around the wall a whole bunch of times." The command was specific. Incorporating the number 7 at the appropriate place in a ritual was supposed to cause some of its power to be passed into the control of the practitioner. The most frequent method of incorporating numbers into magic ritual was by repetition of various parts of the ritual. For example, the following is a spell taken from

a modern book on magic which is supposed to bring back a lover
who has been unfaithful:

> *Who turns from you shall yet be bound*
> *If signs of him may still be found*
> *Within your house—one hair or thread,*
> *Fragment of color, scent or word,*
> *Or any thing that bears his touch—*
> *This spell turns little into much:*
> *Seal the relic in a box*
> *With seven strings tied round for locks,*
> *Each one tight knotted seven times:*
> *Then set on it these seven signs:*
> *Hide it in darkness, out of sight,*
> *Until the next moon's seventh night,*
> *Then send it to the one you seek—*
> *He must return within a week.*[4]

Other numbers frequently used in magic ritual included 3, 5,
and 9. In a chant designed to attract money we find:

> *When the grey owlet has three times hoo'd*
> *When the grinning cat has three times mewed,*
> *When the toad has croaked three times in the wood,*
> *At the red of the moon may this money be good.*[5]

Another example of the use of number in magic comes from the
talismanic magic found in Jewish Kabbalah teaching. Magic
squares are squares which have been subdivided into smaller
squares. Each subdivision contains a number. The numbers within
the various subdivisions are arranged in such a manner that when
the columns, rows, or diagonals are added, the sums are equal. In
Figure 10 we see an example of a common three-by-three magic
square with a total of nine subdivisions. Adding the three columns,
the three rows, and the two diagonals confirms that the sum is
always 15.

Such squares also exist for larger subdivisions. Magic squares
have been known from antiquity, the first reported from Chinese
mathematics. One legend says that the first magic square appeared
on the back of a tortoise which was discovered by the Chinese
Emperor Yu (ca. 2200 B.C.).[6]

| 8 | 1 | 6 |
|---|---|---|
| 3 | 5 | 7 |
| 4 | 9 | 2 |

FIGURE 10. A 3 × 3 magic square. All the rows and columns, as well as the two diagonals, add to the same number, 15.

According to Kabbalah teachings, a specific magic square, called a *kameas*, was associated with each of the following planets: Saturn, Jupiter, Mars, Venus, and Mercury, in addition to both the sun and the moon. The size of these magic squares began with a three-by-three for Saturn and increased up to a nine-by-nine for the moon (Figure 11). The magic squares associated with the heavenly bodies were used to prepare talismans, supposedly transferring the power of the planets, through the numbers within the squares, to some other object or person.

The ancient Hebrew language used letters of the alphabet to stand for both letters and numerals. Hence, any word could be directly translated into a set of numbers corresponding to the individual letters. To prepare a talisman for a specific person or thing, the magician first translated the object's Hebrew name into the corresponding numbers. Lines were then traced on the appropriate *kameas* beginning with the first number in the set, and moving to each succeeding number. This figure was then transcribed inside a circle which had been drawn upon an object, possibly a metal disk or parchment. This disk or parchment became the magic talisman, having acquired power from the planet whose *kameas* was used.

Just why certain planets possess power to influence human events, or why certain magic squares are associated with those heavenly bodies, or how the power of the heavenly bodies is passed to the numbers, and hence to the talisman, we are not informed.

| 4 | 9 | 2 |
|---|---|---|
| 3 | 5 | 7 |
| 8 | 1 | 6 |

Saturn

| 37 | 78 | 29 | 70 | 21 | 62 | 13 | 54 | 5 |
|----|----|----|----|----|----|----|----|----|
| 6 | 38 | 79 | 30 | 71 | 22 | 63 | 14 | 46 |
| 47 | 7 | 39 | 80 | 31 | 72 | 23 | 55 | 15 |
| 16 | 48 | 8 | 40 | 81 | 32 | 64 | 24 | 56 |
| 57 | 17 | 49 | 9 | 41 | 73 | 33 | 65 | 25 |
| 26 | 58 | 18 | 50 | ·1 | 42 | 74 | 34 | 66 |
| 67 | 27 | 59 | 10 | 51 | 2 | 43 | 75 | 35 |
| 36 | 68 | 19 | 60 | 11 | 52 | 3 | 44 | 76 |
| 77 | 28 | 69 | 20 | 61 | 12 | 53 | 4 | 45 |

The Moon

FIGURE 11.  The 3 × 3 magic square for Saturn and the 9 × 9 magic square for the Moon.

## NUMEROLOGY

One ancient field of occult practice was numerology, the belief that numbers influence the lives of people. Those numbers associated with an individual were "discovered" by considering the individual's name and date of birth. Numerology was closely related to gematria, the interpretation of scripture by studying the number equivalents to the Hebrew or Greek words of the scriptures.

The roots of numerology go all the way back to the Pythagoreans and their secret, religious society. We have already mentioned that Pythagoras was profoundly influenced by the fact that certain number ratios of string lengths produced sounds that are in harmony. From this he concluded that the harmonies were produced by the numbers themselves. However, the practitioners of numerology took this idea one step further with the belief that the use of appropriate numbers produces a kind of generalized harmony in nature. This harmony is not restricted to the harmony of tones and their effects upon our ears but extends to objects working or fitting together well. Hence, numbers can be used to achieve harmony in marriage, and in personal and business relations.

A second idea central to numerology was the reincarnation of the spirit. Pythagoras also held to this idea, referring to it as the transmigration of the soul. Numerology, supposedly, helps us understand what our individual spirits are meant to do or accomplish in this lifetime. The key to the relationship between an individual and numbers could be found by translating that person's name and birth date into numbers. These numbers, in turn, were supposed to identify specific characteristics about that individual.

It can be amusing to translate one's name and birth date to see what numbers and characteristics come up. Four ways existed to translate a person's birth date and name into numbers. The first was the *Birth Path* (also called a variety of other names including Life Path, Life Cycle Number, Destiny Number). To compute the number for a birth path was simple: Add the number of the birth month (January = 1 through December = 12) to the day of the month and the year. My date of birth is March 28, 1941. Hence, my birth path is $3 + 28 + 1 + 9 + 4 + 1 = 46$. If the two digits are not the same (as in 33 or 44) then we add these two digits together: $4 + 6 = 10 = 1$. If the two digits are the same, we leave them, for these are special "Master Numbers" and should not be added. Notice that it doesn't matter if we add the various parts of the birth date numbers (month, day, or year numbers) as individual digits or as combined numbers, we always get the same result. Hence, we could form one long number of 3281941 and add these digits to get $28 = 10 = 1$.

Now, what does the birth path number tell me? This number supposedly represents my basic nature. A 1 indicates I'm an individualist, a pioneer, a loner. Considering who I am, I suppose these characteristics, in part, apply to me. Yet, when I consider a list of potential spiritual qualities (including companion, teammate, facilitator, artist, sculptor, aesthetic, worker, steadfast, humanitarian, sage, and counselor) I feel a little cheated. Can't I also be a humanitarian, a sage, an artist? And what about humility? Humorist? Outstanding duck hunter?

Now that I have determined my basic nature (individualist), I can move on to the other three techniques to compute my supposed numerological lessons in life. I translate the letters in my name to numbers and then add these numbers in three different ways. Many systems existed in the ancient world for translating letters to numbers, however I will use one which has survived to the present: the Pythagorean system.

$$a, j, s = 1$$
$$b, t = 2$$
$$c, l, u = 3$$
$$d, m = 4$$
$$e, n, w = 5$$
$$f, o, x = 6$$
$$g, p, y = 7$$
$$h, q, z = 8$$
$$i, r = 9$$
$$k = 11$$
$$v = 22$$

After translating the letters of my name to numbers I can generate three different additional numbers by adding the numbers associated with the vowels, the consonants, and the entire name. Each has a different interpretation. The vowel total gives a number associated with my motivating force. The consonant total is associated with dreams, aspirations, and how those close to me view me. The total name number is associated with my spiritual mission. You may be wondering how my parents were clever enough to pick a name that describes my nature, mission, etc., at birth. As it turns out, I somehow influenced my parents while I was still in the womb to pick for me the name they gave me.

Let's see what happens when I compute my three numbers from my name. You're invited to get a pencil and paper and compute your numbers, too. We all might as well see where we're headed in this life.

Choose the version of your name as it appears on your birth certificate, excluding any Sr., Jr., etc., or titles. I was named after my father, and will, of course, not include my Jr. This means, of course, that my father and I share the same numbers, except for our birth paths.

| C | a | l | v | i | n | | C | l | a | r | e | n | c | e | | C | l | a | w | s | o | n |
|---|---|---|---|---|---|---|---|---|---|---|---|---|---|---|---|---|---|---|---|---|---|---|
| | 1 | | | 9 | | | | | 1 | | 5 | | | 5 | | | | 1 | | | 6 | |
| 3 | | 3 | 22 | | 5 | | 3 | 3 | | 9 | | 5 | 3 | | | 3 | 3 | | 5 | 1 | | 5 |

The first line of numbers are my vowel numbers and total 28 or $2 + 8 = 10 = 1$. This is the same as my birth path. The consonants add to $73 = 7 + 3 = 10 = 1$. What? It seems this is also a 1. Now my total name number is just $28 + 73 = 101$ or 2. Hence, beginning as a loner and individualist, and motivated the same way, my mission in life (represented by 2) is to become a companion, teammate, committee person, facilitator. It could still happen!

What did you discover about yourself when you computed your name and birth date numbers? Did you arrive at special characteristics that meant something to you? To accept numerology one must accept several complex truths on faith.

1. We possess a soul or spirit.
2. This spirit passes from our body upon its death to another body.
3. Our spirits have surviving qualities, e.g., individualism, artistic talent, humanitarianism, etc.

4. Our spiritual natures have been given or have assumed some purpose or mission for their existence.
5. We influence our parents before birth to choose the correct names for us.
6. Each letter in our names is associated in some mystical manner to a number. The three sums of these numbers are associated with our individual spiritual characteristics.
7. By understanding our basic natures and our missions as revealed by numerology, we can bring our lives into harmony.

While a large number of people may be moved to accept any one of the above beliefs, far fewer will accept them all. However, there is still a small group of individuals who believe in numerology. But that is to be expected. There are still people who believe in astrology, and I understand there are flatlanders in England who still believe the earth is flat!

## GEMATRIA

Gematria, or arithmology, is the belief that various writings, especially religious scriptures, contain secret messages that can be found by translating those writings into numbers. In some early languages, including Hebrew and Greek, the letters of the alphabet were also used as numerals. Therefore, the letters of words could also be read as numbers. In fact, entire phrases, sentences, and paragraphs can be translated into numbers. The earliest recorded use of gematria was by the Babylonian King Sargon II who, in the eighth century B.C., built the wall of Khorsabad exactly 16,283 cubits long because that was the numerical value of his name.[7] Those who used gematria generally believed that God gave humans the Hebrew and Greek languages with this double use, and then wrote the scriptures in such a way that the number values of words would have meaning in addition to the obvious meaning of the words. Hence, under this belief, God inserted secret meaning into the Bible which can only be discovered through gematria.

**Table 2.** Greek Numerals and Associated Alphabet

| Units | Tens | Hundreds | Thousands |
|-------|------|----------|-----------|
| 1 = A | 10 = I | 100 = P | 1000 = A |
| 2 = B | 20 = K | 200 = Σ | 2000 = B |
| 3 = Γ | 30 = Λ | 300 = T | 3000 = Γ |
| 4 = Δ | 40 = M | 400 = Y | 4000 = Δ |
| 5 = E | 50 = N | 500 = Φ | 5000 = E |
| 6 = ϝ | 60 = Ξ | 600 = X | 6000 = ϝ |
| 7 = Z | 70 = O | 700 = Ψ | 7000 = Z |
| 8 = H | 80 = Π | 800 = Ω | 8000 = H |
| 9 = Θ | 90 = ϕ | 900 = ϡ | 9000 = Θ |

The connection between gematria and Pythagorean numerology is derived from the ancient Pythagorean doctrine that numbers act both as a blueprint for material existence and that they are the actual constitutes of matter.

The Greek alphabet, with its associated number values, is presented in Table 2. As an example of gematria, we can assign the number values to specific words in the *Bible* and then add the individual numbers in that word to yield one number for the entire word. For example, the Greek word Jesus is ΙΗΣΟΥΣ. Using Table 2 we can substitute in the numerical values for the letters, getting 10 + 8 + 200 + 70 + 400 + 200. Adding these numbers, we arrive at a total numerical value for the Greek name of Jesus of 888. This number, then, is supposed to be a number associated with the name of Jesus. Other phrases in the *Bible* which relate to Jesus can now be decoded and their numbers compared to 888. Such comparisons are supposed to reveal additional sacred messages from God as well as demonstrate the divinity of the Bible. Believers in gematria hold that God used the double meaning of Greek and Hebrew letters to tie the different concepts found within holy scriptures together. This numerical connection supposedly proves the authenticity of the scriptures as inspired by God. The scriptures, when considered in ordinary language, are subject to substantial differences in interpretation. When we then translate the language into just numbers, the possibility for deviation in interpreted meaning increases significantly. If God has hidden messages within the scriptures, God has done a superb job.

An interesting deviation of gematria was a practice among the ancient Hebrews and during the Middle Ages called beasting. In Revelation 13:18 we find:

> Let him that hath understanding count the number of the beast: for it is the number of a man; and his number is Six hundred threescore and six.[8]

From this, Biblical scholars took the number of the devil to be 666. If it was possible to find a combination of letters in your enemy's name that added, through gematria, to be 666, then that enemy would be discredited. This beasting attack was made against a number of powerful individuals, including the Popes of Rome.

## THEY ARE STILL WITH US

Even today we find small groups who are devoted practitioners of magic, numerology, astrology, and gematria. In the ancient world, such beliefs probably appeared to be attractive to many compared to other ancient myths. As a group, modern scientists and mathematicians have rejected these beliefs, objecting to the practitioners' use of the word "science" in describing their activity. A little thought demonstrates that such practices cannot claim to be sciences. All established sciences use a recognized procedure, called the scientific method, which any scientist can rattle off in his or her sleep. This procedure includes several key activities that are missing in the aforementioned belief system.

Science begins with natural observations (phenomena) which are, when first observed, unexplainable. The scientist then proposes a hypothesis that might explain these observations. The hypothesis is then used to predict phenomena that should occur if the hypothesis is a correct explanation. Next, the scientist attempts to observe if the predicted observations actually occur (possibly with an experiment). If the observations do occur, then the predictions are confirmed. If this process is repeated frequently enough, the hypothesis becomes a scientific law. True science deals with statements which can be proven or disproven and relies on the principle of independent verification. If I make a claim, it must be possible for another to objectively verify that claim. Ultimately, utilizing the laws of the science, we are able to make meaningful predictions about the future.

While numerology and gematria may on occasion try to incorporate some of these steps, others are missing. The hypothesis of numerology is that a person's name and birth date, through the use of numerological principles, will reveal that person's basic nature and mission in life. Let us, for the time being, ignore the mission part and concentrate on using numerology to predict an individual's nature. Can we predict basic individual natures by numerologically interpreting that individual's birth data? Is there any statistical evidence that people born on the same day (resulting in the same birth path) share the same nature? No. The verification procedure so essential in science is missing in numerology. Instead, numerology relies on personal anecdotes for confirmation. I see that my birth path is a 1 representing a pioneer, an individualistic individual, and I'm complimented. Yes, that description applies to me (so I let myself believe), therefore numerology must be correct. In truth, the characteristics used by numerology are so general as to apply in some degree to most of us. Hence, the predictive power of numerology is almost nonexistent. If it had true predictive power in the scientific sense, we could compute a birth path and predict if that person was going to become a serial killer or a United States president. Now, that would be a science!

Much the same can be said for gematria. While it may be interesting to study the relationships between the numbers associated with words and the various ideas expressed within the Bible, gematria does not give us a strong tool for making meaningful predictions about the world we live in.

While both numerology and gematria involve the manipulation of numbers, any supposed truths discovered by their practitioners have nothing to do with mathematics proper, but deal with the attributes of people (as in numerology) or God's relationship to the human race (as in gematria). It is true that numbers are involved in both instances, yet numbers are only the vehicles leading toward conditional statements regarding the human condition. In other words, we do not study numerology or gematria to discover or identify a new insight about numbers, for numbers are not the concern of numerology, people are the object of interest. After more than 2000 years of work in the fields of numerology and gematria, their practitioners have not added one new mathematical theorem to our great body of mathematics.

Remembering back to Pythagoras, we recall that he made the logical leap that it was the ratios of numbers that caused the harmony of the vibrating string. Occult practitioners went far beyond this belief and proposed that numbers held some mystical power which controlled individual humans. This idea is discounted by modern science. Yet, is Pythagoras' original idea such an absurd notion, given the modern twists of logic that has produced non-Euclidean geometry, quantum mechanics, and black holes? If mathematical models imitate physical reality, is it too great a leap in faith to say that maybe physical reality is imitating mathematics? Ultimately, we are faced with the question of why— why does mathematics work in the material universe? Both numerology and gematria may be unproductive when viewed as sciences, but the basic impulse which attracts us to them is the same impulse that attracts us to science and applied mathematics. And the basic quandary is not resolved. Why does mathematics work so magnificently as a model to explain our universe? Scientists use mathematical models of the physical world to make claims and predictions about the world. Why should this relationship between model and physical reality exist unless there is some underlying connection? If numbers are only objects of thought, then why are they so wonderfully useful in analyzing the material universe?

## CHAPTER THREE

# SEQUENCES AND SERIES

*No part of Mathematics suffers more from the*

*triviality of its initial presentation to beginners than*

*the great subject of series. . . . the general ideas are*

*never disclosed and thus the examples, which exem-*

*plify nothing, are reduced to silly trivialities.*

ALFRED NORTH WHITEHEAD[1]

## LIMITS—THE MATHEMATICAL HOLY GRAIL

*T*he natural numbers form a sequence, or set of numbers which is ordered in a specific manner. We have already noted that the discovery (or invention) of our counting sequence was certainly one of the greatest of all humankind. And we can use this sequence as a basis for generating even more sophisticated mathematical concepts. From the notion of a sequence we can evolve the concept of limits, one of the most elegant and beautiful ideas in all of mathematics.

Here, it is fun to take a short historical side trip dealing with series. The famous mathematician, Carl Friedrich Gauss (Figure 12), was ten years old when he attended his first day of a particular mathematics class. The instructor, an odd curmudgeon, liked to give himself a rest from lecturing by giving his students a labored problem to occupy their time. As the students finished their work on their small slate boards, they would place the boards upon the instructor's table. When all the slate boards had been stacked on

FIGURE 12.   Carl Friedrich Gauss, 1777–1855.

top of each other, the instructor could see who had finished first
and who was last. On this particular day he gave them the following problem: Add the first 100 natural numbers together.

Gauss immediately wrote an answer upon his slate board and
placed it upon the table. The instructor was incredulous that this
new student could add 100 terms so quickly and he assumed that
answer would be wrong. However, when all the other students had
finally finished their work and placed their slates upon the table,
the instructor turned Gauss' slate over and read the correct answer:
5050. How had Gauss done it?

Gauss noticed that the first term, 1, and the last term, 100, added
together to be 101. Then he realized that the second term, 2, and the
next-to-last term, 99, also added together to be 101. In fact, if he
kept adding pairs of terms in this manner, he would get 50 pairs of

sums, each sum equal to 101. Fifty multiplied by 101 is 5050! We have no record of the words uttered by the shocked teacher when he realized that young Gauss was correct.

Before proceeding we should make a distinction between sequences and series. A number sequence is an ordered set or list of numbers. Frequently we show the members of a set inside brackets { }. The set can contain a finite number of numbers, such as the sequence: {1, 4, 10, 20}. Or it can contain an infinite number of numbers, as in our natural number sequence: {1, 2, 3, 4, 5, . . .}. When we are dealing with an infinite set we use three dots (called an *ellipsis*) after the last number on the right to show that the sequence continues without end. It is important that we specify that the numbers are in a specific order, for it is the very order of the numbers that defines the sequence and gives it meaningful characteristics.

A number series, on the other hand, is the sum of a set of numbers. Hence, every number sequence has its associated number series, i.e., the sum of all the numbers in the sequence. We use special symbols when adding the numbers of a number sequence into a number series. Sometimes we use a large $S$, and if we know the number of terms in the sequence is $n$, we show the sum as $S_n$, meaning that we have added $n$ items. Thus, the sum of the first five numbers in the natural number sequence would be:

$$S_5 = 1 + 2 + 3 + 4 + 5 = 15$$

On other occasions we use the Greek letter sigma or $\Sigma$. Therefore, we can show the above series in several different ways:

$$S_5 = \sum_{i=1}^{5} i = 1 + 2 + 3 + 4 + 5 = 15$$

Below the sigma we have $i = 1$ which tells us the first term we are adding; and above the sigma we have a 5 which shows us the last or fifth term to be added. In other words, we are adding $i$ when $i$ is equal to the numbers 1 through 5. Another example of using the sigma notation is:

$$\sum_{i=1}^{6} i^2 = 1^2 + 2^2 + 3^2 + 4^2 + 5^2 + 6^2$$

Here we show the sum of the first six natural numbers all squared. While this use of the sigma may at first seem awkward, it turns out to be a handy shorthand when we talk about more complex series.

We can add together a finite number of terms in a series and the result will always be a finite sum or number. What happens if we try to add an infinite number of terms together? Does it even make sense to talk about adding an infinite number of numbers together? For a long time many mathematicians claimed that this notion was entirely ridiculous.

One of the first to state his objections was the Greek philosopher, Zeno (489–? B.C.). Almost 2500 years ago Zeno was a member of the Greek Eleatic school, founded by Parmenides of Elea (ca. 475 B.C.). Zeno argued against many of the ideas of the Pythagorean school. One of his arguments involved the absurdity of dividing space into an infinite number of segments and then adding these segments together again. In his argument about the moving arrow, he states that before an arrow can move from the archer to its target, it must first reach the halfway point between them. But, before reaching the halfway point, it must first arrive at the one quarter marker, and so on. This process of subdividing the space between archer and target can go on indefinitely, or, as some say, without end—infinitely many times. If space is infinitely divisible, then, for the arrow to move at all, it must move over an infinity of distances in a finite amount of time—an absurdity!

Zeno's arguments notwithstanding, arrows do manage to reach their targets, and they do this even though the space they move through can, at least in our imaginations, be subdivided an infinite number of times. Thus the sum of all these infinite line segments add to a finite amount—that is, they add together to be the distance from archer to the target. Yet, the notion of adding an infinite number of numbers to get a finite sum continued to bother people from Zeno's time to the present century.

When we say we are adding together an infinity of numbers, we do not mean, of course, that we sit down with pencil and paper and physically create the symbols on paper, doing this for all the infinity of numbers. This process would certainly take an infinite amount of time. What we are really claiming is that numbers can be imagined as added together in our minds, and that this adding process does not actually take any time, i.e., it happens instantaneously. Just as we do not bring to our imaginations every single one of the infinity of natural numbers when we think of the infinite set of all such numbers, we do not bring into our imaginations each and every number we are adding together when adding an infinite series. We have faith that they can somehow come together as a sum.

However, the vagueness of talking about "somehow coming together as a sum" bothers mathematicians. They are worried, and justifiably so, that loose notions will trap them in contradictions later on. Therefore, they avoid talking about adding an infinite number of numbers together by defining what it means for a series to be *unbounded*. When a series (which is really just a sum of numbers) is unbounded, it means that its terms are infinite in number, and that for any number you can think of, no matter how big, we can add enough of the terms of the series together to exceed that number.

This definition is, admittedly, somewhat cluttered. Let's clean things up with a nice example. We can show the series that contains all the natural numbers in the following way:

$$S_\infty = \sum_{i=1}^{\infty} i = 1 + 2 + 3 + 4 + 5 + \ldots$$

We have used both the $S$ symbol and the sigma symbol to represent our series. Notice that in the above equation we have replaced the $n$ above the sigma with the symbol for infinity ($\infty$). This means we are going to begin our series with the first term equal to 1 and then add all the rest of the natural numbers—an infinity of numbers.

By claiming this series is unbounded, I say that no matter how big a number you pick, I can add together a *finite* number of terms from the series to get a sum bigger than your number. For example, you might give me the number 100. "Ah ha!" you exclaim. "Can you beat 100?" If I add together the first 14 terms of the series I get 105.

"Okay," you say, "100 wasn't really a big number. Add enough terms to exceed one million!"

Easy. I add together the first 1414 terms to get the number 1,000,405. The point here is that it doesn't matter how big a number you pick, I can always beat it by just adding enough of the terms from the series. Using this definition of unbounded, I have avoided talking about an infinite number of additions. For every number you give me, I only have to add a finite number of terms to beat it. How did I know how many terms to add to exceed one million? Well, I cheated. I happen to know that if I add together the first $n$ terms from the natural numbers I get the following number:

$$S_n = \sum_{i=1}^{n} i = \frac{n(n+1)}{2}$$

If I need a number, $n$, which will give me a sum exceeding 1,000,000 then I have the following problem.

$$\frac{n(n+1)}{2} > 1,000,000$$

By a little calculating, I find the desired $n$ is 1414, the fewest consecutive numbers beginning with 1 that are greater than 1,000,000. Hence, summing the first 1414 terms in the series gives 1,000,405.

Simply using the definition of unbounded to avoid talking about adding an infinite number of terms does not avoid the problem of infinity. The definition says that for any number you propose, I can find my $n$. That means that for *every* number, or for the infinity of all natural numbers that exist, I can find an $n$ so that the sum of the first $n$ terms exceeds them. Hence, the definition of unbounded still involves the idea of infinity. All we have avoided is talk of making infinite additions.

When we add together all the numbers in an infinite number sequence, we know that the results might turn out to be an unbounded series. Is there such a thing as a bounded infinite series; that is, a series where we add the infinity of terms and get a finite number for a sum?

## GRAPHING SERIES

Our basic number sequence is, of course, the natural number sequence $\{1, 2, 3, 4, \ldots\}$ The associated series for this sequence would be $1 + 2 + 3 + 4 + \ldots$, which we have already determined is unbounded. When a series is unbounded we say that it *diverges* or is a *divergent* series. On the other hand, should we find a series that is not unbounded—that has a finite sum—we will call it a *converging* or *convergent* series.

A simpler unbounded series than the number sequence is the series where each term is just a one, or $1 + 1 + 1 + 1 + \ldots$. We know that the sum of a finite number of terms ($n$) for this series is just $n$, itself, or:

$$\sum_{i=1}^{n} 1_i = 1_1 + 1_2 + 1_3 + \ldots + 1_n = n$$

Therefore, we see that this series is also unbounded. In the above expression we have identified each term with a numbered subscript. We can think of this series as the simplest series because it increases in such a regular fashion, and because each successive term is identical to the preceding one. In Figure 13 we have graphed both the natural number series and our simplest series as the number of terms, $n$, increases. We use a graph here because a graph can give us a kind of mathematical picture of what is happening to our series as we add more and more terms.

We see at once from Figure 13 that the line representing the sum of the natural numbers is above the line for the simple series and will, in fact, stay about that line. This, in turn, tells us that the natural number series increases much faster than our simple series. The simple series has a very special characteristic. Those series that increase faster than this series have individual terms that succes-

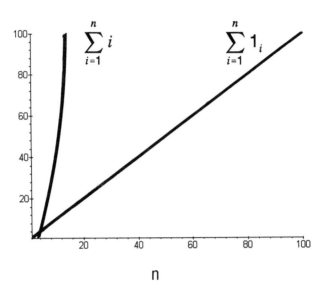

FIGURE 13.  Growth of the natural number series and the simple series.

sively increase in size compared to their predecessors. Those series that increase slower than the simple series have individual terms that decrease in size as the number of terms increases. Therefore, those series with graphs whose lines stay above the simple series graph will always diverge or be unbounded. To find a series that is bounded we are going to have to consider those series that have individual terms which decrease in size, with corresponding graphs that lie below the simple series graph in Figure 13. Can we find such a series?

## THE MAGNIFICENT HARMONIC SERIES

Using the natural numbers we can construct a new series by using the reciprocal of each whole number. Doing so gives us the series $1/1 + 1/2 + 1/3 + 1/4 + \ldots$ or:

$$S = \sum_{n=1}^{\infty} 1/n = 1/1 + 1/2 + 1/3 + \ldots$$

This is called the harmonic series, a series that has been studied since ancient times. On the surface, it would appear that this series should converge because the terms keep getting smaller and smaller. For centuries, many mathematicians believed that the harmonic series did converge. However, Nicole Oresme (1323–1382), the Bishop of Lisieux, France, proved that the harmonic series diverged and was therefore unbounded. He did this in a surprisingly simple way. Let's look at the harmonic series again.

$$S_\infty = 1/1 + 1/2 + 1/3 + 1/4 + 1/5 + 1/6 + 1/7 + 1/8 + 1/9 + \ldots$$

We can group the terms into the following subsets:

$$S_\infty = 1/1 + 1/2 + [1/3 + 1/4] + [1/5 + 1/6 + 1/7 + 1/8] + [1/9 + \ldots$$

Now in each bracketed group we replace each fraction with the smallest fraction in that subset. This substitution yields a new series. However, if we can prove this new series diverges, then we will know the harmonic series must also diverge because the harmonic series is larger than this reconstructed one.

Making the replacements we get:

$$S_\infty = 1/1 + 1/2 + [1/4 + 1/4]$$
$$+ [1/8 + 1/8 + 1/8 + 1/8] + [1/16 + \ldots$$

Now we add the terms together found in each subset.

$$S_\infty = 1/1 + 1/2 + [1/2] + [1/2] + \ldots$$

Hence, every subset is equal to just $1/2$. There are an infinite number of subsets (even though each subset has twice the members as the preceding one). Therefore, there are an infinite number of $1/2$s to be added, and the result will be infinite. Since this smaller series diverges, then the harmonic series, whose terms are individually either equal to or larger than the new terms, must diverge also. We can write this divergence as:

$$\text{Harmonic series } S_\infty = 1/1 + 1/2 + 1/3 + 1/4 + \ldots = \infty$$

What is fascinating about the harmonic series is not just that it diverges, but that it does so very slowly. Let's consider the first 15 terms of the harmonic series in decimal form as shown in Table 3. Notice that by the 15th term we are increasing the sum of the series

**Table 3.** Sums of the First 15 Terms for the Harmonic, Odd, and Prime
Series

| Number of Terms | $\Sigma 1/n$ Harmonic Series | $\Sigma 1/(2n-1)$ Odd Series | $\Sigma 1/p$ Prime Series |
|---|---|---|---|
| 1 | 1.0000 | 1.0000 | 0.5000 |
| 2 | 1.5000 | 1.3333 | 0.8333 |
| 3 | 1.8333 | 1.5333 | 1.0333 |
| 4 | 2.0833 | 1.6761 | 1.1761 |
| 5 | 2.2833 | 1.7873 | 1.2670 |
| 6 | 2.4499 | 1.8782 | 1.3440 |
| 7 | 2.5928 | 1.9551 | 1.4028 |
| 8 | 2.7178 | 2.0218 | 1.4554 |
| 9 | 2.8289 | 2.0806 | 1.4989 |
| 10 | 2.9289 | 2.1332 | 1.5334 |
| 11 | 3.0198 | 2.1808 | 1.5656 |
| 12 | 3.1032 | 2.2243 | 1.5927 |
| 13 | 3.1801 | 2.2643 | 1.6171 |
| 14 | 3.2515 | 2.3013 | 1.6403 |
| 15 | 3.3182 | 2.3358 | 1.6616 |

by only .0666666. . . or 1/15. The sum of these first 15 terms is only
3.3182285. Therefore, we can see the series is increasing in size
slowly, and that the rate of increase continues to slow as the terms
increase in number. Yet, the sum of the series is unbounded, which
means that for any number, say 10,000, we can find how many
terms must be added to exceed 10,000.

Using a computer to actually do the calculations, we discover
it takes an astounding 12,367 terms to add up to just 10! Not 10,000,
but 10. Hence, the harmonic series increases not just slowly, but
incredibly slowly. In fact, to reach 15 we must add the first 1,673,849
terms. With such a slow growth, how can we ever expect the
harmonic series to reach 100 or 1000? Yet, remarkably, it does. Table
4 lists the value of the harmonic series after summing various
numbers of terms.

A graph of how fast the harmonic series grows is shown in
Figure 14. On the horizontal scale we have the number of terms we
have added from 1 up to 1,000,000. On the vertical scale we see the

**Table 4.** Sums for the Harmonic, Odd, and Prime Series

| Number of Terms | $\Sigma 1/n$<br>Harmonic Series | $\Sigma 1/(2n-1)$<br>Odd Series | $\Sigma 1/p$<br>Prime Series |
|---|---|---|---|
| 100 | 5.1873 | 3.2843 | 2.1063 |
| 1000 | 7.4854 | 4.4356 | 2.4574 |
| 10,000 | 9.7876 | 5.5869 | 2.7092 |
| 100,000 | 12.0908 | 6.7385 | 2.9060 |
| 1,000,000 | 14.3838 | 7.8718 | 3.0682 |

corresponding sum of the harmonic series. Notice how flat the line is becoming as the number of terms increases.

To reach 100 we must add approximately $1.5 \times 10^{43}$ terms. That's 15 followed by 42 zeros! To get to 1000 we must add together approximately $1.75 \times 10^{434}$ terms. That's 175 followed by 432 zeros! We can define a relationship between ordinary numbers and the harmonic series (called a function by mathematicians). We let the symbol $H(x)$ represent the number of terms in the harmonic series that must be added to reach $x$. Hence, $H(10) = 12,367$ since it takes 12,367 terms to add to 10, and $H(15) = 1,673,849$.

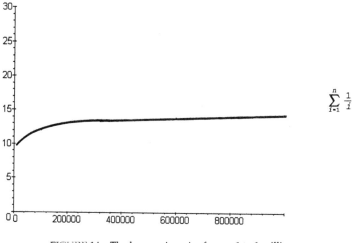

$$\sum_{i=1}^{n} \frac{1}{i}$$

FIGURE 14. The harmonic series for $n = 1$ to 1 million.

We are now in the position to define some really big numbers. Remember in Chapter 1 we defined certain large numbers, including the googol ($10^{100}$) and googolplex ($10^{\text{googol}}$). These numbers become mere pikers when we use them as the $x$ in our harmonic function, for the numbers $H(10^{100})$ and $H(10^{\text{googol}})$ are astronomically larger than $10^{100}$ or $10^{\text{googol}}$.

Since our harmonic series diverges, we still have not found an infinite series that is bounded. Is there another series that grows even slower than the harmonic series? If so, maybe this series will be bounded. Let's consider the reciprocals of only the odd numbers: $1/1 + 1/3 + 1/5 + 1/7 + \ldots$ or:

$$\sum_{n=1}^{\infty} \frac{1}{2n-1} = 1/1 + 1/3 + 1/5 + 1/7 + \ldots$$

This series, which we will call the odd series, is just the harmonic series with the fractions that have even denominators removed. Surely, this series must be bounded for it will grow much slower than the harmonic series. Checking Tables 3 and 4, we see that it does increase much more slowly. The first ten terms add to only 2.133256. At 1,000,000 terms, the odd series is only about half the size of the harmonic series, or 7.871825. But, alas, this series also diverges and is unbounded!

We need a series that grows even slower than either our harmonic or odd series. We can create another series by forming a series of the reciprocals of successive prime numbers. The prime numbers, remember, are those numbers which can only be evenly divided by 1 and themselves. The first prime is 2, followed by 3, 5, 7, and 11. To form the reciprocal prime series we simply take the reciprocal of each of these numbers or:

$$\sum_{p=\text{prime}}^{\infty} \frac{1}{p} = 1/2 + 1/3 + 1/5 + 1/7 + 1/11 + \ldots$$

The series continues on through the infinity of prime numbers. It will have smaller terms than the odd series because the primes thin out as we get to larger and larger numbers. Among the first 1,000,000 numbers there are, of course, 500,000 odd numbers. How-

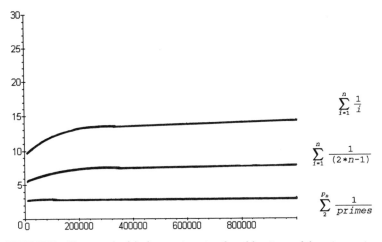

FIGURE 15. The growth of the harmonic series, the odd series, and the prime series through the first million terms.

ever, in the first 1,000,000 numbers there are only 78,499 primes. And they keep getting thinner as we go to numbers beyond 1,000,000. The first 15 terms of the prime series (Table 3) add to 1.6616458, which is considerably less than the 3.3182285 we got by adding the first 15 terms of the harmonic series. In fact, to add up to just 3 we must add over 300,000 terms! After adding together a million terms we still only get the sum of 3.068. In Figure 15 we have included the growth of the sum of the reciprocals of primes along with both the harmonic and odd series. Notice how slowly all three series are growing as the number of terms increases.

With the reciprocals of prime numbers, have we finally found a series that is bounded? No! Amazingly, this series also increases without any end. For any number we choose, no matter how large, we can add enough terms together from the prime series to exceed that number. Just as we did with the harmonic series, we can use the prime series to define a new kind of large number. We designate a function as $P(x)$ which stands for the number of terms needed from the prime series to exceed the number $x$. Hence, $P(3.068) =$ 1,000,000 which means that it takes a million terms to add to 3.068.

A magnificently large number is defined as $P$(googolplex) or: How many terms must be added together from the prime series to exceed a googolplex?

## THE GOLFER'S STORY

Can we ever find a series that is bounded, or was Zeno correct when he said it is folly to believe an infinite number of numbers add to a finite amount? To solve our problem, let's consider the story of a golfer. The day is bright and sunny as we walk to the first tee. This first hole is a 400-yard par 4. In our minds we see our strategy: first a nice drive out 220 yards. Then a four iron shot to the green, followed by a one putt, and we score a birdie!

We put the ball on the tee, position ourselves, and swing. Our shot is right down the middle of the fairway, but a little short, going only 200 yards. We still have 200 yards to the cup. That's okay, with a good wood shot we can make up the distance. We select our three wood, but again the shot is short, falling only 100 yards away, and still 100 yards short of the hole. A little knot grows in our stomach. "Okay," we mutter. "Maybe not a birdie, or even a par, but certainly a bogey!" With only 100 yards remaining to the flag, we select our pitching wedge. Yet, to our horror, this shot, too, is short, going high and falling 50 yards from the flag, and still yards from the edge of the green.

Sweat breaks out on our foreheads. We've already used three shots. At least two or three more are going to be required to get down into the cup. What began as a birdie, now looks like a double bogey—disaster! With a sinking heart, we notice our golfing partners are patiently waiting for us on the green. Using our pitching wedge, we make a little chip-and-run shot at the flag. You can guess what happens. Short again! We're on the green now, but 25 yards from the hole. If we add how far we have moved the golf ball, we get the finite series: 200 yards + 100 yards + 50 yards + 25 yards = 375 yards. With trembling hands, we putt, but it's short, for the ball covers only half the remaining distance to the hole. We putt again, and again we are short by half the distance.

This scenario is repeated, of course, every weekend all across America. However, with some fortitude, and without throwing their clubs into trees or ponds, most golfers manage to somehow sink the ball in a finite number of strokes. We will not allow this for our imaginary game. Each and every time we putt the ball, it rolls only halfway to the cup.

Figure 16 shows the effect of the first few golf shots, and demonstrates that we will, very soon, get exceedingly close to the cup. In fact, after a surprisingly few more putts we will manage to get the ball closer to the edge of the cup than the diameter of a hydrogen atom ($1.74 \times 10^{-10}$ feet). Exactly how many shots to go from 400 yards to under the diameter of a hydrogen atom? Just 43 will do. Yet, the process never stops because with each successive putt we cover only half the remaining distance. Is this series unbounded? No, because it can never exceed the value of 400 yards, for we know the ball always stops some small distance short of the edge of the cup. Hence, the series is bounded by the number 400. By construction, we have finally found our infinite converging series.

In fact, if we think of the distance from the tee to the cup as one fairway distance, then each shot is a fraction of one fairway distance. Using this designation, we have the series: $S_\infty = 1/2$ fairway $+ 1/4$ fairway $+ 1/8$ fairway $+ 1/16$ fairway $+ \dots$ We can show this process with the following infinite series:

$$\sum_{n=1}^{\infty} \frac{1}{2^n} = \frac{1}{2} + \frac{1}{2^2} + \frac{1}{2^3} + \frac{1}{2^4} + \dots$$

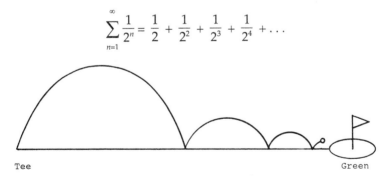

Tee                                                                   Green

FIGURE 16.  Each golf shot travels one half the remaining distance to the pin. The ball will never reach the pin, but will get ever closer to it.

We can consider the number 1 and all numbers larger than 1 as upper bounds on this series. The smallest bound, or 1, has a name: It's called a limit. Mathematicians have a special way of saying that 1 is the limit of this series. They say that 1 is the limit to the series we are building because no matter how small a distance we consider, the golf ball can get closer than this distance to the cup, but at the same time can never reach the cup. Suppose we pick a distance of 1/100,000,000th of a foot. That's only one 1/100,000,000th of a foot—a pretty short distance. Now we can find the number of shots required to get closer to the cup than that distance. In fact, if we take 37 shots we would be only 1/134,217,728th of a foot from the cup. We could, of course, pick an even shorter distance, but if we do, we can always putt enough times to get closer to the cup than that distance. Hence, one fairway is the limit to our game.

Mathematicians like to talk this way because it avoids actually saying they are adding an infinity of numbers together. They avoid talking of infinity altogether and reserve their comments to something like "given any number $\varepsilon$ there exists an $n$ such that $1 - (2^n - 1)/2^n$ is smaller than $\varepsilon$." Loosely translated, it says that given any small number (represented by the Greek letter $\varepsilon$, or epsilon) we can find the number of putts ($n$) when the ball will be closer to the cup than $\varepsilon$.

The idea of a limit is central to all higher mathematics known as analysis. Calculus is built on the idea of limits. Every student who takes a course in calculus begins by studying limits. The idea of a limit is a powerful notion in mathematics. However, even with the use of limits and the special language of mathematicians, we are still forced to consider infinite additions. Even though we have avoided using the term "infinite," we continue to wrap ourselves with the concept. For example, someone might object to our example of the ball getting ever closer to the cup, and to our claim that one fairway is the limit of his golf game. One may claim that after an extraordinary amount of time, the ball might finally equal one fairway and fall into the cup.

Our counterargument is that the fractions we form from the total amount of the distance covered after $n$ shots will always have denominators larger than the numerators. Hence, the total distance traveled by the ball after any finite number of shots will always be less than one.

"Okay," our antagonist responds. "How do you know that after enough days and weeks of putting, some point closer to the ball becomes the limit?" In other words, suppose there exists some infinitesimally small amount less than one fairway that is the real limit to all this putting? Our response is to ask precisely what that infinitesimal amount is, and once known, we can compute how many putts it takes for the ball to get beyond that point. The key here is that we can do this for any small amount chosen less than one complete fairway. Another way to say this is: For *all* numbers, no matter how small, we can putt the ball until it is closer to the cup than that amount. When we use the word "all" we reintroduce infinity—there is just no way to get around the notion. And yet, in a charming sense, it is this very use of infinity that makes the mathematics of series and sequences so much fun. As limited and finite beings, we humans can play with and discover the infinite. Would we want it any other way?

We now have a better idea of what we mean when we say that an infinite series converges to a limit, and we can begin to investigate both converging and diverging sequences and series. We must make one distinction before continuing. We know some infinite series diverge, and we have finally found an infinite series that converges. Yet, there is another case. Some infinite series neither converge to a limit nor diverge.

A series that does not diverge in the sense of being unbounded nor converge to a limit is called an *indefinitely divergent* or, for short, *indefinite* series.[2] Such a series is:

$$S_\infty = 1 - 1 + 1 - 1 + 1 - 1 + \ldots$$

This is an especially interesting series because we can look at it in two different ways. First we group the terms in the following manner:

$$S_\infty = (1 - 1) + (1 - 1) + (1 - 1) + \ldots$$

Now it is obvious that the sum within each pair of parentheses is just zero. Therefore we have the sum of an infinite number of zeros which is just zero. Hence:

$$S_\infty = (1 - 1) + (1 - 1) + \ldots = 0 + 0 + \ldots = 0$$

Yet, we can also group the original series in the following way:

$$S_\infty = 1 + (-1 + 1) + (-1 + 1) + \ldots$$

Here again the terms within the parentheses become zero.

$$S = 1 + 0 + 0 + \ldots = 1$$

Therefore, the series is equal to 1. But we already showed that it was equal to zero! Which is it? You see why this kind of series is called indefinite. It's not really 1 or zero but oscillates back and forth between zero and 1.

## THE FARMER'S PROBLEM

Since recorded history, beginning around 3100 B.C. in the cities of Sumer (present day Iraq), we find references to both sequences and series.[3] However, we can make a strong indirect argument that both series and sequences were known even before the invention of writing. The natural number sequence is, of course, a fundamental sequence that has been known to humankind for at least the last 35,000 years as demonstrated by the baboon bone found in Africa and the wolf bone from Czechoslovakia. Number series must have been considered since the earliest farming days, around 10,000 B.C. in the Fertile Crescent.

Pretend we are farmers living along the Tigris River in Sumer during the year 6001 B.C. One question that we frequently ask is "how much grain should we plant in the spring for an adequate supply of food for the next year, with enough grain left over to plant again for the following year?" Suppose that we know that our seeds will increase threefold because of our farming efforts. Hence, 1/3 of a bushel planted in the spring will yield one whole bushel in the fall. We also know that we need one large bushel of grain to feed our family through the winter. All right, let's plant 1/3 bushel of grain. This yields a full bushel in the fall, and we can survive the winter. The problem is: We don't have any grain to plant the following spring. So, we must plant a little more than 1/3 bushel. How much more? If we want 1/3 bushel of grain next spring for planting, we are going to have to plant 1/9 bushel this spring to generate that 1/3 (since 1/9 bushel will triple to 1/3). Therefore we must plant at least 1/3 + 1/9 of a bushel. So we do. We get 1+1/3 bushel in the fall. We eat the bushel and have 1/3 to plant during the next spring. But wait! That 1/3 bushel planted next spring yields only one bushel the following fall. We won't have anything left over to plant the following spring.

You can now see our conundrum. We must plant 1/3 bushel for the grain we will eat, plus 1/9 bushel for next spring's planting, and 1/27 bushel for the following spring, and 1/81 bushel for the spring after that, and so on. In fact, the correct solution to this problem is the infinite series:

$$\sum_{n=1}^{\infty} \frac{1}{3^n} = \frac{1}{3} + \frac{1}{9} + \frac{1}{27} + \frac{1}{81} + \ldots$$

We may question whether ancient farmers considered solving this infinite series, or even whether they formulated it as an infinite series. In all likelihood, they probably figured that it was best to plant 1/3 bushel to produce enough to eat, and then simply planted another fraction to handle all the rest. However, the farmer's problem is so common that it must have been worked on at various times by individual farmers, elders of farm villages, and even advisors to kings during the thousands of years before the invention of writing.

Surprisingly, an exact solution is available for the kind of infinite series represented by the farmer's problem. To compute this sum we begin with the same series, but only a finite number of terms. We notice that each term is just the preceding term multiplied by 1/3. Therefore, we can rewrite our finite series in the following manner:

$$S_n = \left(\frac{1}{3}\right) + \left(\frac{1}{3}\right) \cdot \left(\frac{1}{3}\right) + \left(\frac{1}{9}\right) \cdot \left(\frac{1}{3}\right) + \left(\frac{1}{27}\right) \cdot \left(\frac{1}{3}\right) + \ldots + \left(\frac{1}{3^n}\right)$$

This equation can be rewritten in the following manner:

$$S_n = \sum_{i=1}^{n} \frac{1}{3^i} = \frac{1}{3^1} + \frac{1}{3^2} + \frac{1}{3^3} + \ldots + \frac{1}{3^n}$$

A series where each succeeding term is the previous term multiplied by a constant is called a *geometric* series or a *geometric progression*. The constant term we are multiplying by is 1/3 and is called the common *ratio*. Hence, to get the next term in the series we only have to multiply the previous one by 1/3. We can rewrite the above series in the following way:

Line 1     $S_n = (1/3)^1 + (1/3)^2 + (1/3)^3 + \ldots + (1/3)^n$

Now, we simply multiply both the left and righthand sides of the above equation by the common ratio of 1/3 to get:

$$(1/3)S_n = (1/3)(1/3)^1 + (1/3)(1/3)^2 + \ldots + (1/3)(1/3)^n$$

or multiplying the terms together on the right we get:

Line 2     $(1/3)S_n = (1/3)^2 + (1/3)^3 + (1/3)^4 + \ldots + (1/3)^{n+1}$

Next we subtract the equation on Line 2 from Line 1. Notice that all the in-between terms on the right cancel out to yield:

$$S_n - (1/3)S_n = (1/3)^1 - (1/3)^{n+1}$$

We can factor $S_n$ out of the left side.

$$S_n(1 - 1/3) = (1/3)^1 - (1/3)^{n+1}$$

Now we simply divide both sides by $(1 - 1/3)$:

$$S_n = \frac{\left(\frac{1}{3}\right)^1 - \left(\frac{1}{3}\right)^{n+1}}{1 - \frac{1}{3}}$$

We next separate the numerators into two different fractions:

$$S_n = \frac{\frac{1}{3}}{1 - \frac{1}{3}} - \frac{\left(\frac{1}{3}\right)^{n+1}}{1 - \frac{1}{3}}$$

Notice that the first term on the right depends only on the ratio $r$, or $1/3$, and the first term, again $1/3$. The second term on the right is dependent on the number of terms we sum together. We can see at once that the numerator of this term, $(1/3)^{n+1}$, will get smaller and smaller as $n$ grows larger. Now suppose we let the series go to infinity. The number $n$ will grow infinitely large causing the numerator to decrease to zero. Hence, when $n$ goes to infinity, the sum becomes the first term on the right or:

$$S_\infty = (1/3)/(1 - 1/3) = (1/3)/(2/3) = 3/6 \text{ or } 1/2$$

Therefore, the correct solution to the farmer's problem is to plant exactly $1/2$ bushel of grain.

We can restate the solution to the farmer's problem in general terms. Suppose we have the following geometrical progression:

$$S_\infty = a + ar + ar^2 + ar^3 + ar^4 + \ldots$$

We apply the same method of solution and find the general solution to be:

$$S_\infty = a/(1 - r)$$

Now we don't care if we're talking about a ratio of $1/3$, $1/2$, or $1/10$. In each case we can find the exact solution. The only condition we must insist on is that $r$ be less than 1. If $r$ is 1 or greater, then the series will not converge but diverge. If the ratio is less than 1, the geometrical progression will always converge.

The techniques we used were not beyond the talents of the early farming civilizations of the Fertile Crescent. However, we did employ the notion of an infinity of numbers, and we talked rather loosely of a number $n$ growing to infinity. Such notions may well have been too advanced for our early farming ancestors. Yet, the farmer's problem is so universal, it would be incredulous to think that early farmers never considered the sums of finite series. Such series are more relevant to real-life situations than they seem on first consideration.

## ARITHMETIC AND GEOMETRIC SEQUENCES AND SERIES

The farmer's problem illustrated one of two simple types of sequences and series: the arithmetic and geometric progressions.

The arithmetic progression is a series or sequence where each succeeding term is arrived at by adding a fixed amount to its predecessor. For example, natural numbers are an arithmetic progression because each term is simply its predecessor plus 1. The amount we add to the predecessor each time is called the common difference, which we will designate as $d$. If we let $a_1$ be the first term, then the nth term, $a_n$, is just: $a_n = a_1 + (n - 1)d$, where $d$ is the difference and $n$ is the number of terms.

Therefore, the tenth term in the sequence of natural numbers is simply:

$$a_{10} = 1 + (10 - 1) \cdot 1 = 1 + 9 = 10$$

But, we knew this already. Let's look at a more intriguing arithmetic sequence: 7, 11, 15, 19, 23, . . . . In this sequence the first term is 7 and the difference is 4. What would the 100th term be? We simply substitute the proper amounts into our equation and get

$$a_{100} = 7 + (100 - 1) \cdot 4 = 7 + 396 = 403$$

Our 100th term is 403, and we learned this without having to add a hundred different numbers together.

The other type of special sequence, the geometric sequence, is of interest because of the many places we encounter it. A modern use of it is in compounded interest or compounded economic growth. As mentioned earlier, in a geometric sequence we don't add the same term to each number, but we multiply each number by the same amount, the ratio. A simple geometric sequence is: 1, 2, 4, 8, 16, 32, 64, . . . . In this sequence we start with the number 1 and then begin to double, or multiply by 2. Hence, our ratio is 2. This doubling sequence crops up in many places. The ancient Egyptians used a doubling procedure to carry out their multiplication. For example, to multiply two numbers, say $11 \times 13$, they would generate the following list.

Multiplying $11 \times 13$

| 11 | 1 | \ |
|----|---|---|
| 22 | 2 |   |
| 44 | 4 | \ |
| 88 | 8 | \ |

In the left column we begin with 11 and then simply double it three times. In the right column we have the multiplier beginning with 1 next to 11. The Egyptian scribe would place a hash mark next to each number in the right column to indicate those numbers adding to 13. Hence we see a hash mark next to the numbers 1, 4, and 8 since 13 = 1 + 4 + 8. He would then add the corresponding numbers in the left column together or 11 + 44 + 88 = 143, which is, in fact, equal to 11 × 13. This method was so simple and reliable for multiplication that it was used for centuries after the Egyptians and became known as the Russian Peasant Method of multiplication.

Another example of the doubling sequence is in the design of modern computers. Computers are based on a binary system, and therefore the memory modules are multiples of 2. Hence a kilobyte of memory is not 1000 bytes, but is really 1024 bytes. This is because the sequence representing the powers of two is 2, 4, 8, 16, 32, 64, 128, 256, 512, 1024, etc.

The formula for the $n$th term in a geometric sequence is:

$$a_n = a_1 r^{n-1}$$

If we wish to learn the tenth term in the above doubling sequence we simply substitute 1 for $a_1$, 2 for the ratio, and 10 for $n$.

$$a_{10} = 1 \cdot 2^{10-1} = 2^9 = 1024$$

One kind of infinite sequence is of special interest: the sequence whose terms continue to diminish to ever smaller amounts in such a way that we can always move along the sequence to find a number as close to zero as desired. These sequences are called *evanescent* sequences. An example of a geometrical progression which is also evanescent is:

$$1/2, 1/4, 1/8, 1/16, \ldots$$

No matter how small of a number you give, I can move along the above sequence until I find a smaller value. Does this mean that all evanescent sequences have converging series? Not necessarily.

## ANCIENT EVIDENCE

Number sequences and their associated series may seem to be exotic mathematical inventions with little application to the real world. Certainly in our daily lives we use the number sequence to count things, but we seldom use a series, and we certainly never use an infinite sequence or series. (Who has the time for all that adding?) Yet, if we take a careful look, we will see that sequences and series are very old, and are associated with fundamental human activities.

The oldest existing writings on mathematics contain references to number series. On an ancient Babylonian clay tablet we find a list of the values associated with the fraction of the moon's disc that is illuminated over a complete lunar cycle. The first five days are recorded in a geometrical progression with the succeeding values given in an arithmetic progression.[4]

The Rhind Papyrus mentioned in Chapter 1 was written around 1650 B.C. and may have been copied from an earlier papyrus scroll dating around 1900 B.C. The Rhind Papyrus contains 84 problems. Problem 40 states that 100 loaves of bread are to be divided among five men in such a way that the men receiving the two smallest shares receive a total that is $1/7$ of the shares of the other three men. From the way the problem is stated, it is evident that the shares should be in arithmetic progression. The question is to find the difference between the shares, i.e., the common difference between the terms of the progression. The solution to this problem is the arithmetic progression:[5]

$$10/6 + 65/6 + 120/6 + 175/6 + 230/6 = 600/6 = 100$$

Notice that the first two shares total $75/6$ which is exactly one seventh of $525/6$, the sum of the other three shares. The common difference between each term is $55/6$ or 9 and $1/6$ loaves of bread. This demonstrates that the ancient Egyptians knew of sequences and series. In fact, the way they used fractions would suggest the concept of a series to them since all Egyptian fractions, except $2/3$, were unit fractions with a numerator equal to the number 1. Hence, fractions such as $11/15$ had to be written as the sum of unit fractions, e.g., $11/15 = 1/2 + 1/5 + 1/30$.

## CHAPTER FOUR

# THE FAMILY OF NUMBERS

*So if man's wit be wandering, let him study the*

*mathematics; for in demonstrations, if his wit be*

*called away never so little, he must begin again.*

FRANCIS BACON

*ESSAYS*[1]

## ONE POTATO, TWO POTATO

*I*n order to fully appreciate the full range of numbers we are going to encounter in our odyssey, it is helpful to review the various kinds of numbers defined by mathematics. We begin, of course, with the natural numbers: 1, 2, 3, 4, . . .

Most certainly, the natural numbers were the first numbers encountered by humans. However, as soon as farming began, new problems made demands the natural numbers could not fulfill. Hence, we discovered (invented?) the positive fractions: 1/2, 3/2, 1/1, 17/91, 3/1, . . . Notice that we have defined both 1/1 and 3/1 as fractions, when, in fact, they are the natural numbers 1 and 3. This is a nice convention which allows us to include all the natural numbers within the set of all fractions. In this way we can say that fractions include all numbers of the form $n/m$ where both $n$ and $m$ are natural numbers.

To illustrate our numbers we will relate them to the number line. When using a number line we assign numbers to specific points in relation to their size. This approach allows us to visualize impor-

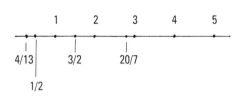

FIGURE 17. The number line showing the location of various whole numbers and fractions.

tant features about numbers as we define them. Figure 17 shows a number line in which we have assigned 1, 2, 3, 4, and 5 to specific points. For convenience, we space these points equal distances apart. We have also shown the points assigned to 1/2, 4/13, 3/2, and 20/7 as examples of positive fractions.

It is interesting to note that two other kinds of numbers in common use, decimals and percentages, are really fractions. A percentage is the numerator of a fraction that has a denominator of 100. For example 37% is really the fraction 37/100. A decimal is really the sum of a series of fractions. For example, the decimal 31.47 is the sum of the following three fractions: 31.47 = 31/1 + 4/10 + 7/100.

Decimals come in three flavors: terminating decimals (e.g., 3.5), nonterminating, periodic decimals (e.g., 0.13131313. . .), and nonterminating, nonperiodic decimals (e.g., 5.973821. . .). By nonterminating, we mean that the digits to the right of the decimal in the decimal number continue on for an infinite number of places. By periodic we mean that a fixed number of digits repeat indefinitely, beginning somewhere to the right of the decimal point. If the decimal is both nonterminating and periodic, then the repeating digits continue to repeat without end. A simple example is the fraction 1/3 which is equal to the decimal, 0.33333 . . . .

Both terminating and nonterminating, periodic decimals can be rewritten as either a finite sum of fractions or a convergent infinite sum of fractions. (The third kind of decimal, the nonterminating, nonperiodic decimal, we will leave until later.) Since these two kinds of decimals can be represented by series of fractions, we have a nice correspondence between fractions and decimals involving finite and infinite series. For example, consider again the fraction 1/3. If we use long division to divide 3 into 1, we begin with a decimal of the form 0.3333 . . . and the process of division never ends, because we repeatedly get a remainder in the division. This means that we have the following identity:

$$\sum_{n=1}^{\infty} \frac{3}{10^n} = \frac{3}{10^1} + \frac{3}{10^2} + \frac{3}{10^3} + \frac{3}{10^4} + \cdots = \frac{1}{3}$$

Here we have the kind of bounded infinite series we searched so long for, and it is equal to the fraction 1/3. In fact, we can expand this idea even further. Since every fraction has a corresponding decimal number, we can see that every fraction is equal to a corresponding bounded series. If the fraction is terminating, such as 1/5 = 0.2, then the corresponding series will be finite. If the decimal is a nonterminating, periodic decimal then the corresponding series will be bounded and infinite. Hence, every fraction is equal to a bounded, infinite, or finite series.

I like to show my beginning algebra class a little trick with infinite repeating decimals. We can form the decimal 0.99999. . . where the ellipsis indicates the 9s go on forever. I then claim that this decimal is exactly equal to the number 1. I usually get a few students who scoff at my claim, so I show them a proof. First we multiply the decimal by 10 getting 9.99999 . . . which is the original decimal with the decimal point shifted one place to the right. We subtract the original decimal from this new one:

$$10X = 9.99999\ldots$$

$$\underline{-X = -.99999\ldots}$$
$$9X = 9.00000\ldots$$

Now we divide both sides by 9 to get $X = 1$. I did this proof in class one day only to have a young student become very upset and agitated. Her intuition told her that .99999 . . . was some infinitesimal amount less than 1, and here I was doing some kind of witchcraft to contradict her belief. I made an effort to review the theory of decimals with the class to ease her discomfort. I'm not sure I succeeded, for she left class shaking her head and muttering to herself.

When we show a point on our number line, such as in Figure 17, we show a small black round dot. A true point, however, has no length or height. But such a point cannot be seen. Therefore, the dots we show on paper are only meant to help us visualize approximately where we intend our points to be. Since points have no length, they are really markers of position, not small line segments nor little round dots that have some magnitude. We must keep this in mind while developing our theory of numbers, and also remember that points are not equal to numbers. We are only assigning numbers to points (positions on the line) in order to show their relative magnitude in relation to each other.

Both the Babylonians and Egyptians understood how to compute with positive fractions, and seemed to treat them as real numbers along with the natural numbers. The Greeks, on the other hand, didn't consider fractions as real numbers. What we call fractions, they called ratios between natural numbers.

The next additions we want to make to our number line, the negative numbers and zero, evolved over many centuries. Negative numbers may have been used first by the Chinese. By 300 B.C. they were using counting boards that employed counting rods to make calculations. Sometime after this, they began using red rods to signify positive numbers and black rods to signify negative numbers. It is believed that the black rods were used to show debt, a popular use for negative numbers today.

One of the chief pressures exerted on any theory of numbers is for the numbers to satisfy various kinds of equations, equations being symbolic statements of mathematical operations. If we add any two natural numbers, we always get another natural number as an answer. This is convenient for it insures that the operation of addition will always produce another number, and not something strange and undefined. Hence we have: $X = 5 + 4$ as a symbolic representation for adding the natural numbers 5 and 4. The "solution" of course is 9 ($X = 9$). If we multiply any two natural numbers, we always get the same kind of thing as a result—that is, we always get another natural number. Hence we have: $X = 4.5$ and a solution which is $X = 20$. This ability to get from our arithmetic operation the same kind of thing we begin with is known in mathematics as *closure*. We say that the natural numbers are closed under both addition and multiplication.

When we shift to division and subtraction, this closure breaks down. When we have an equation such as: $X = 3 \div 2$ we do not get a natural number as a result, but a fraction, instead. By including the fractions within our collection of numbers, we assure ourselves of closure for the numbers under the operation of division, for if we divide any natural number by any other natural number, we get a fraction as an answer (we include the natural numbers as a special kind of fraction).

When we consider subtraction, the closure of fractions breaks down. For example, $X = 5 - 7$ is an equation that does not yield a natural number nor a positive fraction for an answer. For most ancient peoples, such an equation would simply have no solution. What sense can be made of subtracting seven geese from five geese? What is minus 2 geese? Even though the Chinese were able to use negative rods with their counting boards to represent debt, they did not allow negative numbers to be solutions to equations.

The first to allow for true negative numbers were probably the Hindu mathematicians of India. *Brahmagupta* (ca. 628) developed a sound theory to handle negative numbers, allowing them to be the solution to equations. Not only did Brahmagupta use negative numbers, but he also included another number which was not accepted in the West for many more centuries—zero. Now, with the addition of zero and the negative numbers to the collection of numbers, it was possible to add, subtract, multiply, and divide natural numbers and fractions, and always be assured of getting a number as an answer—with one exception—we cannot divide by zero.

If we consider the collection of numbers that includes the natural numbers, negative whole numbers, and zero, then we have the *integers* which we can represent as $\{\ldots, -3, -2, -1, 0, 1, 2, 3, \ldots\}$.

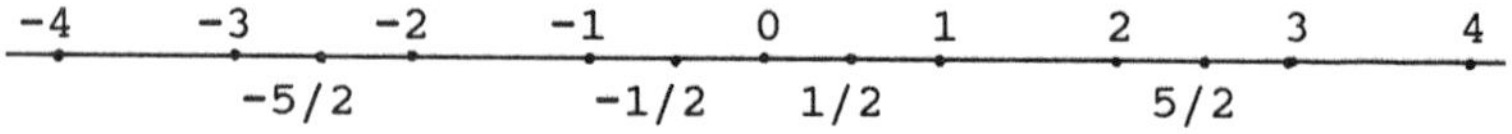

FIGURE 18.   The number line showing the location of various positive and negative numbers.

Now we add all the fractions, both positive and negative, and together these numbers are called the *rational* numbers.

One might think that we have all the numbers we will ever need. Shown in Figure 18 are examples of positive and negative whole numbers, positive and negative fractions, and zero. The set of rational numbers is closed under all four arithmetic operations except for division by zero. What more could we ever need?

## THOSE CURIOUS GREEKS

Although the Greeks did not accept fractions, negative numbers, or zero as numbers, their deductive mathematics allowed them to discover an entirely new kind of number. This discovery turned out to be more troublesome for their theories than beneficial.

Influenced by the Pythagoreans, many Greeks believed that numbers were the atoms that constituted the material world. According to the Pythagorean legend of creation, at the beginning of the universe there existed "The One," "The Limited," a monad without differentiation or extension. Surrounding this monad was the unlimited, which was the principle of extension (space). Somehow the unlimited separated the monad into individual atomic numbers. These numbers, in turn, organized themselves geometrically to form simple shapes, which in turn became the four elements, earth, air, fire, and water.

An important characteristic of this process was harmony, which was identified with the correct ratios between whole numbers. Hence, the geometric constructs must have dimensions that are in whole number ratios to each other. To show a simplified example, we can consider a right triangle whose sides are in the ratios of 3:4:5. The lengths of the sides of this triangle are 3 units, 4 units, and 5

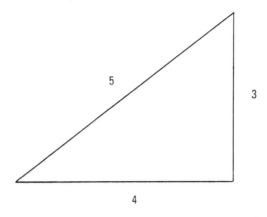

FIGURE 19. A right triangle whose sides are in the proportions of 3, 4, and 5.

units (Figure 19). Now let's look at a square with sides equal to one unit each (Figure 20). We can now ask: If the sides are of one unit, then what is the length of the diagonal? According to Pythagorean cosmology, it should be some whole number ratio, or in modern terms, a fraction of some kind.

Some pesky Pythagorean simply could not let things alone. He went and proved that for a unit square the length of the diagonal could not be a fraction. The proof is beautiful and involves only

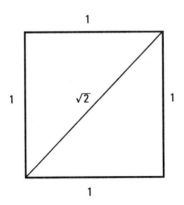

FIGURE 20. The unit square and its diagonal of $\sqrt{2}$.

elementary ideas. We use the Pythagorean theorem, a theorem that was credited to Pythagoras himself by many ancient historians, but a theorem we know was used by both the Egyptians and Babylonians. Whether these ancient peoples possessed a proof is not known. The Pythagorean theorem says that for a right triangle, the sum of the squares of the two legs equals the square of the hypotenuse. Applying this theorem to Figure 20, we see that the hypotenuse is the diagonal of the square while the two legs are the sides of the square. Using $D$ to represent the length of the diagonal we get: $D^2 = 1^2 + 1^2$ or $D^2 = 1 + 1 = 2$. To find the value of $D$ we must find the square root of both sides. Hence, we have $D = \sqrt{2}$. Therefore, we must find a number which, when multiplied by itself, gives us the number 2, or $\sqrt{2} \cdot \sqrt{2} = 2$.

We will use a kind of proof known to the ancient Greeks as *reductio ad absurdum*. With this proof we assume the opposite of what we want to prove. Then we show that this assumption leads to a contradiction. If the opposite of what we wish to prove is false, then the statement we are trying to prove must be true.

We begin by supposing the opposite of what we want to prove in this case: that $\sqrt{2}$ is represented by some ratio of whole numbers, which we designate as $p/q$. If we can find a fraction $p/q = \sqrt{2}$, then we know the length of the diagonal is equal to this fraction. We will assume that $p$ and $q$ have no common factors (for if they did, we would simply cancel them out and get a new $p$ and $q$ that satisfy our needs.) If $p$ and $q$ have no common factor, then they both cannot be even numbers, for if they were both even, then they would both contain a 2 as a factor.

Step 1: (our assumption)                    $p/q = \sqrt{2}$
Step 2: (we square both sides)              $p^2/q^2 = 2$
Step 3: (we multiply each side by q$^2$)    $p^2 = 2q^2$
   Since $p^2$ is equal to $2 \times q^2$, then $p^2$ must be an even number. This takes a moment of reflection, so stop and think about it. If a number is equal to 2 times another number, then the original number must be an even number.

Step 4: $p^2$ is even

If $p^2$ is even, then $p$ must be even. Again, this takes a moment of reflection. If we take an odd number and square it, do we ever get an even number? No. Therefore, since $p^2$ is even, then $p$ is also even.

Step 5: $p$ is even

Step 6: $q$ is odd (if $p$ is even then $q$ must be odd)

If $p$ is even then $q$ must be odd, because we agreed at the beginning that $p$ and $q$ had no common factors.

Step 7: let $p = 2r$ (since $p$ is even)

Step 8: $(2r)^2 = 2q^2$ or $4r^2 = 2q^2$

Here we have simply substituted the $2r$ for $p$ in Step 3.

Step 9: $2r^2 = q^2$ (dividing both sides by 2)

This equation says that $q^2$ is an even number because it is equal to 2 times another number.

Step 10: $q^2$ is even

Step 11: $q$ is even

This is the contradiction we are after, for we have shown that $q$ must be both even and odd. Therefore, when we assume that some fraction exists that is equal to $\sqrt{2}$ we get a contradiction. This tells us that no fraction exists equal in length to $\sqrt{2}$. The Greeks called $\sqrt{2}$ an *incommensurable* length because it could not be represented as the ratio of whole numbers.

This discovery was such a scandal to the Pythagoreans that all were sworn to secrecy. Someone, possibly Hippasus of Metapontum, revealed the secret to outsiders. One legend says that Hippasus was drowned for his treason to the Pythagorean Order.[2] In any case, the Greeks had stumbled onto an entirely new kind of number. Because their theory of numbers did not allow them to consider anything as a number except natural numbers and their ratios, the incommensurable length of the diagonal had to be separated from arithmetic. Because of this discovery, the Greek mathematicians divided mathematics into the disciplines of geometry, which was the study of lines and points, and arithmetic—the study of numbers. All of Euclid's theorems were defined in terms of geometric

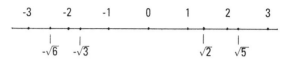

FIGURE 21. The number line showing the positions of the radical numbers, √2, −√3, √5, and −√6.

objects, and not numbers. The division between arithmetic and geometry lasted for 2000 years until René Descartes pulled the two disciplines together again with his analytic geometry.

What shall we call our new kind of number represented by √2? Are there other examples? As it turns out, and was proven by the Greeks, any square root of a whole number that is not a perfect square is one of these incommensurable numbers. Therefore, √3, √5, √6, √7, and √8 are all incommensurable numbers. The only kinds of numbers we leave out are √4 = 2, √9 = 3, and any other number which is a perfect square. This means there are an infinite number of these new numbers. We will call such numbers *radical* numbers. On the number line in Figure 21 we have added several of these new numbers to show their approximate location.

If we now consider all the numbers we have discovered, including the rational numbers and our radical numbers, our new and bigger set of numbers is called the *algebraic* numbers. The reason they are called algebraic leads to an interesting insight. Consider the equation: $A \cdot X + B = 0$ where $X$ is our unknown number, and both $A$ and $B$ are known, integer numbers (we exclude the possibility that $A = 0$). We can solve for $X$ in terms of $A$ and $B$ to get: $X = -B/A$. The value for $X$ will always turn out to be a fraction, zero, or integer. Therefore, we can say that the solution for $X$ will always be a rational number. This kind of equation, $A \cdot X + B = 0$, is called a *linear* equation because the $X$ is not raised to a power greater than 1. Therefore, we can say that all linear equations have solutions that are rational numbers.

Now let's consider a more complex equation of the form: $A \cdot X^2 - B = 0$. Solving for $X$ we get:

$$X = \sqrt{\frac{B}{A}}$$

Here we see that the solution is in the form of a radical. Now, if $A$ and $B$ are both positive numbers, the value of the fraction under the radical sign will be positive. (We do not want to consider at this point what happens when the value under the radical is negative.) If both $A$ and $B$ are perfect squares or their division leads to a perfect square, then we will get a rational number for a solution. However, for all those cases where $A$ and $B$ are not perfect squares, the solution will be one of our new numbers, a radical number. From all this we can say that if there are solutions of the equation $A \cdot X^2 - B = 0$, then these solutions will be algebraic numbers, i.e., the solutions will be either integers, fractions, or radicals.

We can generalize this idea in the following manner. Equations that consist of terms involving an unknown, such as $x$, raised to positive whole numbers (exponent) and multiplied by constants are called polynomial equations. Such equations are classified by the largest exponent found on the unknowns. The general form of the first degree polynomial is simply $A \cdot x + B = 0$. The general form for the second through fourth degree polynomials are as follows:

Second degree   $A \cdot x^2 + B \cdot x + C = 0$
Third degree    $A \cdot x^3 + B \cdot x^2 + C \cdot x + D = 0$
Fourth degree   $A \cdot x^4 + B \cdot x^3 + C \cdot x^2 + D \cdot x + E = 0$

We usually write polynomials in standard form which means we begin on the left with the term containing the largest exponent of $x$ and proceed in descending order. Using this technique we can write the general form for a polynomial of the $n$th degree as:

$$a_0 X^n + a_1 X^{n-1} + a_2 X^{n-2} + \ldots + a_{n-1} X + a_n = 0$$

where the different $a$s are all integers, and $n$ is always a positive integer. For thousands of years mathematicians have been interested in the solutions to polynomials, that is, the values of $x$ that make the equation true. We now know that solutions to this kind of polynomial equation (if they exist for our number line) will always be algebraic numbers. This is a powerful piece of mathematics. It tells us that any equation in the above form, if it has solutions on our number line, will have solutions that are algebraic numbers. It would appear that expanding our collection of numbers to include the radicals has allowed us to account for the solution of many more equations.

Remember when we said that three kinds of decimals existed? We explained that terminating and nonterminating, periodic decimals were equal to fractions, but we ignored what nonterminating, nonperiodic fractions were. Now we have found examples of such decimals. The radicals are infinite, nonperiodic decimals. Therefore, we now know that numbers such as $\sqrt{2}$ and $\sqrt{3}$ have infinite decimal expressions that never become periodic.

Have we accounted for all the points on the number line by adding all the radical numbers to our rational numbers, yielding the algebraic numbers? If you were an ant, an ant so thin you had no width, could you walk along the number line and not fall through? Is every single point accounted for—is every point assigned some algebraic number?

## OF COURSE, THEN THERE WAS π

We know that π is the ratio of the diameter of a circle to its circumference. It has the approximate value of 3.14159265.... What kind of number is π? Is it a fraction? No. Early civilizations used fractions to approximate π. For example, one Egyptian estimate was $(16/9)^2$, which yields a value of 3.16049..., a value only 6/10 of 1% in error.[3] One estimate by the Chinese mathematician Tsu Ch'ung-chih (430–501) was 355/113, which is even better, accurate to seven decimal places.[4] However, no fraction exactly equals π. Therefore, it must be some other kind of number. Is it algebraic— some kind of radical? If π is algebraic, then it will be the solution to some polynomial that has integer coefficients.

If π is not algebraic, then it must be some other kind of number, a number that transcends the algebraic numbers. It would be a transcendental number. Do such transcendental numbers even exist? This question was not answered until the middle of the 19th century. In 1844 Joseph Liouville (1809–1882) proved that transcendental numbers existed by actually constructing examples. He used an infinite series to do this.

In order to understand how he wrote his transcendental numbers, we must examine a special kind of mathematical notation called a factorial. The sign for a factorial is !. When we write N! we mean all the natural numbers up to and including N all multiplied together. Therefore, 1! = 1; 2! = 1·2 = 2; 3! = 1·2·3 = 6. You can see at once that factorials increase very quickly in size, since 10! = 1·2·3·4·5·6·7·8·9·10 = 3,628,800. Using factorial notation we can now show an example of one of Liouville's transcendental numbers.

$$\sum_{n=1}^{\infty} \frac{1}{10^{n!}} = \frac{1}{10} + \frac{1}{10^2} + \frac{1}{10^6} + \frac{1}{10^{24}} + \cdots$$

If we write the first few digits out in decimal form we get:

$$\sum_{n=1}^{\infty} \frac{1}{10^{n!}} = 0.110001000000000000000001000 \ldots$$

The decimal expansion of this number contains zeros everywhere except those locations $n!$ ($n$ factorial) from the right of the decimal point, where $n$ are consecutive numbers beginning with 1. Hence, we have a 1 at positions 1! = 1, 2! = 2, 3! = 6, 4! = 24, etc. The next digit 1 will appear at 5! = 120 or 120 places right of the decimal. Any number of the form

$$\sum_{n=1}^{\infty} \frac{A}{10^{n!}}$$

where $A$ is a constant is a Liouville number.

Liouville proved that such numbers cannot be the solutions of polynomial equations with integer coefficients, and hence were not algebraic. This means they are not any kind of number we have studied so far.

Even though Liouville proved that transcendental numbers exist, it was still 38 years later, in 1882, that C.L.F. Lindemann (1852–1939) proved that $\pi$ was not the solution to any polynomial equation and therefore was a transcendental number. Now we can appreciate how unique π really is. It is not a whole number, fraction, or even an algebraic number, but one of a class of strange new numbers that have only been known for the last 150 years. What kind of decimal representations do we have for transcendental numbers? These numbers, like radicals, have infinite, nonterminating decimals. Therefore, we know that the decimal expansion of π will never reveal an infinitely repeating pattern.

The last question we want to consider regarding these new numbers is: How many are there? If there are just a few, like π and the Liouville numbers, maybe we can sweep them under the carpet and not worry about them.

## JUST HOW MANY ARE MANY?

We have already seen one great mathematical breakthrough in the last century with the discovery of transcendental numbers. We now pause to consider another wondrous discovery by the brilliant mathematician Georg Cantor (1845–1918). We know, of course, that an infinite number of natural numbers exist. For most of recorded history, humans have assumed that an infinite collection of things was just infinite—it went on forever and was impossible to count. That was it—infinite was just that—infinite. Georg Cantor proved that there are actually different sizes to infinity. This idea was so bizarre to many mathematicians that Cantor was attacked for his ideas during much of his career.

One use of numbers is to tell us how many objects are in a collection. From our previous discussion we recall that mathematicians refer to a collection of numbers as a set, and the individual numbers are elements in the set. Finite sets contain just a finite number of elements. We call the number of elements within a set the set's cardinal number. While it is obvious that all finite sets have a corresponding cardinal number, what about infinite sets?

What Cantor proved was that infinite sets do have corresponding infinite cardinal numbers. The set consisting of all natural numbers represents the smallest size of infinity, i.e., the cardinal number for the natural numbers is the smallest, infinite cardinal number. When we deal with finite numbers, certain laws seem to work, and we tend to take these laws for granted. For example, if we add a positive number $B$ to the positive number $A$, we get a different number that is larger than both. Hence, $A + B = C$, and $C > A$ and $C > B$. We do not have to be told that if $2 + 7 = 9$, that 9 is larger than both 2 and 7. When we move on to infinite collections, this rule breaks down. For example, let $\aleph$ be the cardinal number of an infinite set.[5] What is $\aleph + 1$? The result of this addition is not a new, larger infinite set, but just the set $\aleph$, again. Hence, $\aleph + 1 = \aleph$. In fact, let $A$ be any finite number and we have $\aleph + A = \aleph$. This seems to contradict our intuition, but that is because we are dealing with infinite, and not finite sets. We can go even further. In the mathematics of infinite sets we have: $\aleph + \aleph = 2\aleph = \aleph$. Hence, by doubling the number of elements in $\aleph$ we still get $\aleph$ back—that is, we get a new set that has the same cardinal number, $\aleph$.

The strangeness of infinite sets was recognized long before Cantor when the famous scientist and mathematician, Galileo Galilei (1564–1642) made the strange assertion that there appeared to be the same number of positive square integers as there are positive integers. He did this by pointing out that we can map the positive integers onto the positive square integers in the following way:

| 1 | 2 | 3 | 4 | 5 | 6 | 7 | 8 | 9 | 10 | 11 |
|---|---|---|---|---|---|---|---|---|----|----|
| ↕ | ↕ | ↕ | ↕ | ↕ | ↕ | ↕ | ↕ | ↕ | ↕ | ↕ |
| 1 | 4 | 9 | 16 | 25 | 36 | 49 | 64 | 81 | 100 | 121 |

When we have completed this mapping for all the numbers (in our imagination, of course) then we see that for every number there is a corresponding square, and for every square there is a corresponding number.

What this means is that there exists many infinite sets that are the same size as the set of natural numbers, while at the same time they are a subset of the natural numbers. In fact, the natural numbers can be a subset of another infinite set, which is also the same size (has the same cardinal number) as the natural number set. In a situation that seems so muddled, Cantor brought clarity. He defined the smallest infinite set as having the cardinal number $\aleph_0$ (called aleph-null) and called the set a countable set. All infinite sets that are countable can be put into a one-to-one correspondence with the natural numbers, and their size is $\aleph_0$. Are all infinite sets the size of $\aleph_0$?

The natural numbers are, of course, a countable set. What about all integers, including the negative numbers and zero? This infinite set is also countable. What about all the fractions? Certainly, it is intuitively obvious that if we look at the collection of all fractions we must have a larger set than $\aleph_0$. This looks obvious because there are an infinite number of fractions between every two whole numbers.

The genius of Cantor was that he proved the set of all fractions is a countable set, thus, the cardinal number for the rational numbers is just $\aleph_0$. Now, what about if we add the radicals and consider the set of all algebraic numbers? Certainly we must have a larger infinity here! Not so, proved Cantor. All algebraic numbers can be put into a one-to-one mapping with the natural numbers. Hence, the cardinal number for all algebraic numbers is just $\aleph_0$.

We come now to our final question. What about the transcendental numbers? We have only learned of two such numbers, the example of a Liouville number, and $\pi$. Could we possibly expect that the transcendental numbers form a larger infinity than all the algebraic numbers? Again Cantor showed his genius. He proved that the transcendental numbers formed a larger infinite set, designated as C (for continuum). It is impossible to make a one-to-one mapping of the transcendental numbers with the natural numbers. Why? There are just too many of them. These transcendental numbers, numbers so strange and exotic that they were only discovered 150 years ago, and so rare in common usage that the only familiar one is $\pi$, actually represent the great bulk of all numbers on the number line. If we consider the algebraic numbers, we can devise a scheme to make an infinitely long list of these numbers, and the list will include every one. However, because there are so many more transcendental numbers we could never make a list of them all, even if we allowed the list to be infinitely long. If we imagine the number line with only the algebraic numbers in place, and the transcendental numbers left out, then such a line would have more holes than number points. Once before we considered dumping all the natural numbers into a great barrel and randomly drawing one out, asking what kind of number we might get. We can repeat the exercise with the numbers on our number line. Suppose we could randomly select a point on the number line and then look at its associated number. What kind of number would it be? We can answer this question. It would be a transcendental number! The vast infinity of transcendental numbers is so much greater than the infinity of algebraic numbers that the odds, for all practical purposes, are zero that we would randomly pick an algebraic number.

## SO MANY NUMBERS!

We have covered the territory we wanted. We took the number line and found all the numbers that fill this line up.

Natural numbers            1, 2, 3, 4, . . .
Zero                       0
Negative whole numbers     −1, −2, −3, . . .
            *All of the above = Integers*

Fractions                      $1/2, 3/7, -2/9, 23/6, . . .$
            *All of the above = Rational numbers*

Radicals                       $\sqrt{2}, \sqrt{5}, \sqrt{14}, . . .$
            *All of the above = Algebraic numbers*

Transcendental numbers     $\pi, \Sigma 1/10^{n!}$
            *All of the above = Real numbers*

Therefore, all of the numbers on the number line are designated as the real numbers, or sometimes just the reals. We also know that the real numbers form a set that is infinitely larger than the algebraic numbers. Algebraic numbers are represented by the cardinal number $\aleph_0$ and the transcendental numbers by the cardinal number C. Do cardinal numbers exist that are larger than aleph-null and C? Of course, but that is another story.

## CHAPTER FIVE

# STORY FOR A RICH MAN

*The sciences, even the best—mathematics and*

*astronomy—are like sportsmen, who seize whatever*

*prey offers, even without being able to*

*make any use of it.*

RALPH WALDO EMERSON

*REPRESENTATIVE MAN*[1]

## EULER'S WONDERFUL SUM

$\mathcal{W}$e have looked at several series that keep growing without any bound. We have also looked at geometrical series, with ratios less than 1, that converge to a value of $a_1/(1-r)$ where $a_1$ is the first term. What other series exist that converge to an interesting value? If we take a moment, we should see that the discovery of an infinite number of numbers summing to one value must be ranked as one of the most outstanding achievements of humankind. The whole idea flies in the face of intuition. How can we take an infinite number of numbers, add them up, yet end up with a finite sum?

Nicole Oresme (1323?–1382), the mathematician who first proved that the harmonic series diverged, also proved that the following series converged to 2.[2]

$$\sum_{n=1}^{\infty} \frac{n}{2^n} = \frac{1}{2} + \frac{2}{2^2} + \frac{3}{2^3} + \ldots + \frac{n}{2^n} + \ldots = 2$$

He also proved the following series converged to 4/3.

$$\sum_{n=1}^{\infty} \frac{n \cdot 3}{4^n} = \frac{1 \cdot 3}{4} + \frac{2 \cdot 3}{16} + \frac{3 \cdot 3}{64} + \ldots + \frac{n \cdot 3}{4^n} + \ldots = \frac{4}{3}$$

So far, we have only studied series that converge to rational limits, that is, limits that are whole numbers or fractions. However, one of the most remarkable discoveries in mathematics involves a frequently studied series: the sum of the reciprocals of square numbers or:

$$\sum_{n=1}^{\infty} \frac{1}{n^2} = \frac{1}{1^2} + \frac{1}{2^2} + \frac{1}{3^2} + \frac{1}{4^2} + \ldots$$

Many believed $\Sigma\, 1/n^2$ converged since we can compare its terms with the previous series, $\Sigma n/2^n = 2$, and see that the first terms of the former series (excluding the first term) are smaller than the terms of the latter series. Hence, if $\Sigma\, n/2^n$ converged to 2, then $\Sigma\, 1/n^2$ might converge to something less than 2. However, once we get beyond the tenth terms of the two series, those of the series $\Sigma\, 1/n^2$ become larger. Jacob Bernoulli (1654–1705) of the famous Bernoulli family proved that $\Sigma\, 1/n^2$ did converge to a finite value, but to what value he couldn't say. Leonhard Euler found the answer.

Leonhard Euler (1707–1783) was one of the greatest and most productive mathematicians who ever lived. Since we will run into him on numerous occasions, it is worth the effort to become acquainted with this remarkable man. Euler (Figure 22) attended the University of Basel, receiving his bachelor's degree at 15 and his master's degree at 16. At 18 he published his first mathematical paper, and only seven years later, at 25, published a two-volume text on mechanics. Tragically, he lost the sight of his right eye just three years later. Near the age of 60, he became completely blind. Yet, while blind he published over 400 mathematical papers, most of which he dictated to a servant untrained in mathematics. He was a dedicated husband to his wife and a loving father to his 13 children. Euler was probably the most prolific mathematician to ever live, publishing enough math to fill 90 volumes.

FIGURE 22.  Leonhard Euler, 1707–1783.

Euler was fascinated by infinite series and continued the work
in this field begun by men such as Isaac Newton, Gottfried Leibniz,
and the Bernoullis. In 1736 he discovered the limit to the infinite
series, $\Sigma 1/n^2$. He did it by doing some rather ingenious mathematics using trigonometric functions that proved the series summed
to exactly $\pi^2/6$. How can this be? How can the sum of an infinite
series be connected to the ratio of the circumference of a circle to its
diameter? This demonstrates one of the most startling charac-

teristics of mathematics—the interconnectedness of, seemingly, unrelated ideas. Once we become immersed in the study of mathematics, we suddenly stumble onto many of these strange and wondrous connections.

Euler went on to show the convergence of the following two series, both of which have limits involving $\pi$.

$$\sum_{n=1}^{\infty} \frac{1}{(2n-1)^2} = \frac{1}{1^2} + \frac{1}{3^2} + \frac{1}{5^2} + \frac{1}{7^2} + \cdots = \frac{\pi^2}{8}$$

$$\sum_{n=1}^{\infty} \frac{(-1)^{n+1}}{n^2} = \frac{1}{1^2} - \frac{1}{2^2} + \frac{1}{3^2} - \frac{1}{4^2} + \frac{1}{5^2} - \cdots = \frac{\pi^2}{12}$$

The occurrence of $\pi$ in the above series is not a fluke. For reasons not always apparent, $\pi$ shows up in many places that seem to be entirely unrelated to the ratio of a circle's circumference to its diameter. This is not a cosmological problem concerning the universe we live in, but rather a logical problem. The appearance of $\pi$ in seemingly unrelated places is a puzzlement of the rational world, i.e., the world of ideas, and not the world of atoms and galaxies. All of the objects of mathematics are objects of thought, and hence objects of the rational world. Yet, the rational world seems to contain objects whose relationship to each other must be discovered. As soon as I learn the definition of $\pi$ as being the ratio of the circumference of a circle to its diameter, I do not have the knowledge of the occurrence of this ratio throughout the rational universe. Why is this? Is not all thought a creation of human mental activity? If I can define $\pi$, why can't I see at once its many relationships to other mathematical objects?

But $\pi$ is not unique. Other mathematical objects exist that are related in mysterious ways to infinite series and to $\pi$. It is as if there existed some great landscape of meta-mathematics, and we are only seeing the peaks of mathematical mountains above valley fog. That we see $\pi$ on numerous peaks is strange and wonderful to us. Yet, if the dense fog of our ignorance would only dissipate, we could then see the entire landscape of interconnected rational truths, and our understanding would enter a new dimension. Is this concealing fog related to the limitations of our human minds? Does this fog drift away for some alien species living at the center of our galaxy whose IQ is measured in the tens of thousands? (I can't stand this—I must know more. Onward!) We have studied several sequences of numbers that grow without bound. The sequence of prime numbers (2, 3, 5, 7, 11, . . .) is infinite, and therefore, never ends. Yet, the primes thin out as the numbers increase in size. Of the first 100 numbers, 25 (25%) are prime. Of the first 1000 numbers, only 163 are prime (16.3%). The first numbers are rich in primes and, as the natural numbers get larger, fewer primes are found. Do the primes continue to thin out as the natural numbers get larger? Yes.

We have also talked of the harmonic series that grows ever so slowly, yet also increases without bound. We have looked at both the series represented by reciprocals of the odd numbers and the reciprocals of the prime numbers, both series increasing beyond any limit. We can now ask: How fast are these sequences and series increasing, how dense are the primes as the natural numbers increase, and how do we describe how fast the harmonic series grows? To understand how such mathematical creatures behave, we must relate the story of The Rich Man.

## LONG AGO, IN A VILLAGE FAR AWAY

When we studied one particular type of geometric series, we discovered that infinite series are related to an ancient problem common to all farmers, i.e., how much should a farmer plant to insure enough to eat and still have enough left over to plant the following spring? We now encounter a problem that may be just as old. Before farming, humans were hunter–gatherers, generally nomads living off the wild animals they hunted and the wild plants they foraged. Could such people accumulate wealth? That is, before domesticated animals and farmland, could the hunter–gatherers acquire enough material possessions to be considered wealthy in their neighbor's eyes? Possibly. But certainly by the time our ancestors began to farm, some of them soon acquired an abundance of possessions, and hence became wealthy.

Farming made the acquisition of meaningful wealth possible. A single individual or family could accumulate large land holdings, great stores of crops, and many farm animals. Once such wealth is acquired, it is only natural for those possessing it to ask: How can I make this wealth make me more wealth? Hence, the idea of lending property and charging something for its use was born. Suddenly we have interest to be made from wealth.

Therefore, a very early mathematical problem for rich farmers or merchants would be: How do I compute interest? Simple interest is a percent of the principal loaned. We can show this as: Interest = Principal (the original loan) multiplied by the rate of interest, or $I = P \cdot R$. Simple interest may have sufficed for millennia, yet at some time a clever person must have stumbled onto the idea of compounded interest—charging not only for the original loan, but also charging for some of the interest owed. Now, how are we to compute compounded interest? Surprisingly, the equation for compounded interest isn't all that difficult. We have the interest rate, $R$, the original loan, $P$, and now the number of times we are going to compound the interest, $n$. The formula is $A = P(1 + R/n)^n$, where $A$ is the final total owed to us, including both the original loan and the interest.

Let's see how this formula works. If we are a rich farmer living in the Fertile Crescent around the year 6001 B.C. we may want to lend a poor farmer a bushel of grain which he will use for seed to grow a crop. We, of course, expect our original bushel back next fall, plus our interest. Pretend that we are charging 100% interest for the use of our grain for that one year. (Remember this is 6001 B.C. and we're just making up the rules for lending money. Who knows if 100% was a high interest rate for grain back then?) In terms of simple interest we would earn: $I = (1 \text{ bushel})(100\%) = (1 \text{ bushel})(1.00) = 1$ bushel of grain. Hence, the farmer to whom we lend the bushel of grain owes us two bushels at the end of the year: one to repay the bushel he borrowed, and one bushel in interest. We, of course, are very pleased. Notice that in our formula, we replaced the 100% with its decimal equivalent or 1.00. This is standard procedure.

But after lending out the bushel of grain for a number of years and only receiving two bushels, we decide that it would be nice to make more money on our wealth. Isn't this always the case? Once we have people who manage through hard work and cleverness to grow wealthy, some of those same people become greedy. We will assume for this example that we are such greedy people. We could, of course, simply increase the interest rate. But suppose the king of the land had issued a decree that 100% was the maximum interest to be charged on a loan. How are we to get around this limit? To avoid exceeding the king's limit, we decide to lend the grain for six months at 50%. Then we will lend the same bushel plus any interest earned for an additional six months. Hence, we have stayed within our 100% per year limit, but when we apply the formula we discover we've made just a smidgin more.

$A = P(1 + R/n)^n$ is our formula. Now substitute in our values ($A = 1$, $R = 1.00$, $n = 2$):

$$A = (1 \text{ bushel})(1 + 1.00/2)^2 = 1(1 + .5)^2 = (1.5)^2 = 2.25$$

Hey, this is great! Instead of getting just two bushels back, we received the two bushels plus an additional quarter of a bushel. This compound interest is swell.

But, greed being what it is, we are soon bored with our two and a quarter bushels at the end of the year. We want MORE! What if we increased the number of periods that we compounded? Will that increase our effective interest even more? Let's try it. We decide to compound *every three months* or four times a year ($R = .25$ and $n = 4$). The poor farmer who borrows the grain every year knows something isn't right, but we assure him we are staying within the king's edict of no more than 100% interest on loans.

Now our formula becomes: $A = (1 \text{ bushel})(1 + 1.00/4)^4$. What will this produce?

$$A = 1(1 + .25)^4 = (1.25)^4 = 2.44 \text{ bushels}$$

Now we're getting back almost a half bushel of grain more than the simple interest formula. We're definitely on to something. We increase the compounding periods again to every month, or 12 times per year.

$$A = (1 \text{ bushel})(1 + 1.00/12)^{12} = (1.08333. . .)^{12} = 2.61 \text{ bushels}$$

You can see where this is leading. As we become greedier and greedier, we keep increasing the number of compounding periods. Can we increase the number of periods so that the total loan to be paid back continues to increase to any amount we may desire? In other words, is there no limit to the grain we can earn off one bushel by just increasing the compounding periods while the basic interest rate stays at 100%? Let's go for broke and compound the interest DAILY! That's 365 compounding periods.

$$A = (1)(1 + 1.00/365)^{365} = 2.715 \text{ bushels}$$

Not too much of an increase over our 2.61 bushels, but an increase, nevertheless.

We can compound every hour for a year or 8760 compounding periods. However, this increase in compounding periods yields only 2.718 bushels of grain. Our scheme seems to be breaking down. Maybe there exists some limit to how much one bushel will yield no matter how many compounding periods we break the year into. This would mean that the expression $(1 + 1/n)^n$ has an upper bound or limit as $n$ grows larger and larger.

In fact, such a limit exists and is called simply $e$. Symbolically we show this limit as:

$$\lim_{n \to \infty} \left(1 + \frac{1}{n}\right)^n = e$$

The use of $e$ for this limit was introduced by Euler in the 18th century, and has become one of the most important numbers in mathematics along with $\pi$. Our $e$, like $\pi$, is a transcendental number. Hence, the decimal expansion of $e$ yields an infinite, nonrepeating decimal. Therefore, we can't specify its value exactly with either a fraction or finite decimal. Its approximate value is 2.718281828459045. . . . To remember so many digits of $e$ is remarkably simple. Watch what happens when I break the digits into groups: $e = 2.7\ 1828\ 1828\ 45\ 90\ 45. . . .$ After the 2.7 we have the 1828 repeated twice, then 45, followed by twice 45 (90) and then 45 again. We can remember 1828 as the year Joseph Henry discovered electric induction. Now we never need to dash about, frantically trying to look up the value of *e* ever again.

What does all this talk of *e* being a limit imply for our little scheme to get rich from lending out one bushel of grain? It means that no matter how often we increase the number of compounding periods, we will never get back as much as, or more than 2.718281828459045. . . bushels. In other words, if we were to compound infinitely often, or *continuously*, our result would be exactly *e*.

The applications for using the number *e* are nearly countless throughout mathematics. For example, we can consider interest problems when the interest rate is any amount and not just 100% because the limit of the expression $(1 + r/n)^n$ where *r* is the interest rate is just $e^r$ or *e* raised to the *r* power. In fact, the expression we use to define *e*, $(1 + 1/n)^n$, describes many other kinds of changes besides the accumulation of interest. Populations of animals and humans seem to grow in the same manner as interest grows, and such growth patterns are called exponential growth. The reverse process is called exponential decay. For example, dead bodies lose heat exponentially, and therefore *e* can be used in an appropriate equation to determine how long individuals have been dead.

If we were to compute $(1 + 1/n)^n$ for each successive natural number *n* beginning with 1, we would generate a sequence of numbers whose limit was, of course, *e*. But we can also define *e* in terms of an infinite series in the following way:

$$e = 1 + \frac{1}{1} + \frac{1}{1 \cdot 2} + \frac{1}{1 \cdot 2 \cdot 3} + \frac{1}{1 \cdot 2 \cdot 3 \cdot 4} + \cdots$$

Each successive denominator is just the previous denominator multiplied by the next number in the natural number sequence. We know such a product as a factorial written as *n*! Hence, our infinite series for *e* becomes:

$$\sum_{n=0}^{\infty} \frac{1}{n!} = \frac{1}{0!} + \frac{1}{1!} + \frac{1}{2!} + \frac{1}{3!} + \frac{1}{4!} + \cdots = e$$

This series converges quickly; adding just the first seven terms gives us *e* accurate to the thousandths place. A closely related series can also be used to calculate a limit on a compound interest problem, such as our rich farmer experienced. Suppose the interest rate on a loan was 22%. What would the limit on this loan be if the number of compounding periods kept increasing in number? We know that the problem reduces to the limit on the expression:

$$(1 + .22/n)^n = e^{.22}$$

But how do we compute $e^{.22}$? For that we have the nice series:

$$e^x = 1 + \frac{x}{1!} + \frac{x^2}{2!} + \frac{x^3}{3!} + \frac{x^4}{4!} + \cdots$$

Substituting in the appropriate values for $x$ we get:

$$e^{.22} = 1 + \frac{.22}{1} + \frac{(.22)^2}{2} + \frac{(.22)^3}{6} + \frac{(.22)^4}{24} + \cdots$$

After adding only four terms we get 1.246. . . an answer accurate to the thousandths.

Once Euler found the connection between infinite series, $\pi$, and $e$, there was no stopping him, and he found numerous other identities. One such discovery is the following identity which many mathematicians consider to be the most elegant mathematical expression ever discovered:

$$e^{\pi\sqrt{-1}} + 1 = 0$$

This expression needs a little explanation for the uninitiated. We now know what $e$ is, for it is just the limit to the sequence generated by $(1 + 1/n)^n$, and we know what $\pi$ is, for it is the ratio of the circle's circumference to its diameter. But what in the world is $\sqrt{-1}$? From our elementary algebra, we remember that any number, either positive or negative, when multiplied by itself gives a positive result. Now we are asked for a number, which when multiplied by itself, gives us a negative 1. Impossible! In fact, there is no number in the set of real numbers that satisfies this need. To find such a number as $\sqrt{-1}$ we must go to another, expanded group of numbers called *complex* numbers. We will take a closer look at the complex numbers later. For now, we are satisfied that such a number as $\sqrt{-1}$ exists.

Now the simple expression discovered by Euler, $e^{\pi\sqrt{-1}} + 1 = 0$, can be considered in the following light: It contains all the most important elements to be found in the foundations of mathematics. It contains the number 1, the beginning of our natural number sequence. It contains the operation of addition, which can be used to define all the other natural numbers and the other three mathematical operations. It contains zero, a concept that required millennia to evolve for the human race. It also contains both $e$ and $\pi$, the two most important transcendental constants we know. It contains the unit number for defining the complex numbers, $\sqrt{-1}$. And last, but certainly not least, the equation, since it contains $e$, relates the infinite sequence, or:

$$\lim_{n \to \infty} \left(1 + \frac{1}{n}\right)^n$$

to the other fundamental mathematical ideas. What an astounding number of mathematical concepts to be rolled up into just one expression!

What we must not overlook is the wonderful connection between $\pi$ and $e$. We were amazed when we discovered that $\pi$ was associated with infinite series, now it turns out that $\pi$ is also associated in some mysterious way with our limit $e$. But we have only scratched the surface, for the number of beautiful connections between various mathematical entities will continue to reveal themselves as we continue our odyssey.

Another beautiful expression involving a limit that connects not only $\pi$ and $e$, but also radicals and factorials is:

$$\lim_{n \to \infty} \frac{e^n n!}{n^n \cdot \sqrt{n}} = \sqrt{2\pi}$$

## LOGARITHM: A DANCE FOR LUMBERJACKS?

We did not embark on this strange odyssey only to define $e$ and show that it is connected to $\pi$. What we really want to do is demonstrate that $e$ is connected in a deep and fundamental way to the very heart of all mathematics. Using $e$ we will define the tools needed to understand many of the series and sequences we are encountering.

In the first year of algebra, students are exposed to the theory of logarithms, which, in most cases, sends their hearts into a great flutter, for anything with such a long and strange name must be very hard. In fact, logarithms are quite simple. They are exponents. "Now," you say, "what the heck are exponents?"

Exponents are the small numbers we write as superscripts to other numbers. They simply tell us how many times the original number is to be multiplied by itself. Hence, we have $A \cdot A = A^2$ and $A \cdot A \cdot A = A^3$. Thus, an exponent represents repeated multiplication of a number called a base. In our last example, $A$ is the base, 3 is the exponent, while the whole thing, $A^3$, is called the power. Now watch what happens when we multiply two bases together that are equal.

$$A^2 \cdot A^3 = (A \cdot A) \cdot (A \cdot A \cdot A) = A^5 = A^{2+3}$$

Here we see that to multiply the two powers of $A$ together on the left, all we really had to do was add their two exponents. This turns a multiplication problem into an addition problem. Because it is frequently much harder to multiply two numbers than to add them, we can use this characteristic of exponents to solve some rather sticky computational problems. Another nice characteristic of exponents is the following.

$$(A^2)^3 = (A \cdot A)^3 = (A \cdot A) \cdot (A \cdot A) \cdot (A \cdot A) = A^6 = A^{2 \cdot 3}$$

Here we see that to raise a number $A$ that is already squared to another power, we just multiply the powers. In the above example we multiplied the 2 and 3 together. This will also turn out to be a nice characteristic of exponents and lead to simplification with the use of logarithms.

During the Renaissance, the European nations invested a great deal of energy and wealth into exploring the rest of the world, and, in so doing, established new trade routes. This required their sailors to make long and dangerous voyages in small, wooden ships. It was of paramount importance for the ship's navigator to know where they were on the open oceans. If the ship became lost, it could mean disaster. In order to find the ship's position, the navigator had to have a very accurate clock. Knowing the date and exact time, he could check the positions of the stars and compute the ship's latitude and longitude. However, to get from the observation of the stars to knowing where they were, the navigator was required to make ponderous, and very accurate, calculations. A slight error in calculations could cause him to misjudge their true position by hundreds of miles.

Now we have seen that we can use exponents to change a multiplication problem, which can be tedious and difficult, into a simpler addition problem. Our hero of the moment is John Napier, who realized this great need to simplify calculations. He worked for 20 years to invent a system called logarithms, which is based on this idea of exponents.[3]

John Napier was born in 1550 in Scotland, and later became Baron of Murchiston. Although he was not a professional mathematician, he had a strong interest in simplifying calculations. His first paper on this subject, *Mirifici logarithmorum canonis descriptio* ("A Description of the Marvelous Rule of Logarithms") was published in 1614; his second, *Mirifici logarithmorum canonis constructio*, appeared in 1619, two years after his death. Napier's work was quickly refined by others and adopted as a method to simplify difficult calculations in the sciences, trade, and exploration.

Just what is this "marvelous rule of logarithms" of Napier's? We have already seen that when multiplying two powers with the same base, we can simply add the two exponents. Hence, for $10000 \cdot 1000 = 10^4 \cdot 10^3 = 10^{4+3} = 10^7$. If we consider the exponents of the two 10s, we see that $4 + 3 = 7$. When we write a logarithm, which is just an exponent, we must designate the base and the number we get when using the exponent with that base. We do this in the following manner:

$$\log_b x = y \text{ is equivalent to } b^y = x$$

Hence, the logarithm is really just an exponent that we have designated as $y$. The base is $b$, and when we use the exponent $y$ with $b$ we get the value $x$. Now consider our original example. We have $10000 = 10^4$ or $\log_{10} 10000 = 4$. The second statement is read as "the log to the base ten of ten thousand is four." This means that to get 10,000 we must raise 10 to the fourth power. We can define exponents to be any real number, and are not restricted to using only integers. Therefore, having an exponent that is a decimal or fraction is no problem. Logarithms really show their power when the problems are a little more sophisticated and a lot more difficult.

Now it would be senseless to calculate various equivalent exponents for all the infinity of bases, so it has become convenient to use just two numbers as bases: 10 and $e$. Yes, that's right, we're back to $e$ again. It is such an important constant that we use it as one of our two logarithmic bases. When we use logarithms with base 10 we write them as simply $\log A$ without the base indicated in a subscript. These are called *common* logarithms. When we use $e$ as a base we write the logarithm as $\ln x$ instead of $\log_e x$ and understand the base is $e$. Using base $e$ gives us *natural* logarithms.

To give an example of using logarithms to solve difficult problems we can return to one of our original problems: computing interest on loans. Suppose it is July 3, 1776, and a group of men are trying to write some political document in Philadelphia. Because of all the noise outside, they can't hear themselves think. They ask your ancestor, since he owns a stable nearby, to spread a little straw about the road. Your ancestor complies, and in gratitude they give him a promissory note for $1.00 to earn interest of 12% per year, compounded yearly. You have just found your ancestor's note and decide to redeem it. What is it worth in 1996 after 220 years?

The formula is relatively easy. $A = P(1 + I)^t$ where $A$ is the amortized amount (the total we owe at the end of the loan), $P$ is the principal, $I$ is the interest rate and $t$ is the number of years. This yields

$$A = P(1 + I)^t = \$1 \cdot (1 + .12)^{220} = (1.12)^{220}$$

Now all we have to figure out is some way to multiply 1.12 by itself a total of 220 times. We could sit down with pencil and paper and begin to multiply, but with so many calculations we're bound to make a mistake, to say nothing of being reluctant to start the lengthy, tedious process. Here is where we can save considerable labor by employing logarithms. What we will do is use the logarithm of $(1.12)^{220}$, simplify that logarithm, and then change the logarithm back into a normal number, which will be our answer. In using logarithms, we rely on tables that have already been computed by others, the earliest of such tables having been constructed by Napier himself. Therefore, to find the logarithm of a number we look that number up in a logarithm table. Today, with modern hand calculators, we don't have to drag around some large tome of tables, but can get the required logarithm directly from the calculator.

| | |
|---|---|
| Step 1: the original problem | $(1.12)^{220}$ |
| Step 2: change to a logarithm | $\log (1.12)^{220}$ |
| Step 3: reduce the exponent | $220 \cdot \log 1.12$ |
| Step 4: look up logarithm | $220 \cdot (0.049218)$ |
| Step 5: multiply | $10.827964$ |
| Step 6: change back to normal number | $10^{10.827964}$ |

We change this last expression into a product, $10^{.827964} \times 10^{10}$. The first number in the product we look up in our logarithm tables (or get out of our hand calculators).

$$A = 6.729224035 \times 10^{10} = \$67{,}292{,}240{,}350$$

Hence, the government of the United States owes you sixty-seven billion, two hundred and ninety-two million, two hundred and forty thousand, three hundred and fifty dollars for that one dollar of hay purchased in 1776. No wonder our national debt keeps increasing.

What was a most tedious problem to begin with became a rather easy problem using logarithms. That is why logarithmic tables were so eagerly adopted for solving navigation problems and for doing calculations in astronomy, both areas where great accuracy was desired.

## WHAT DO LOGARITHMS LOOK LIKE?

We can compare powers, i.e., bases with exponents, with the exponents, themselves. For example, we can compare the sequence

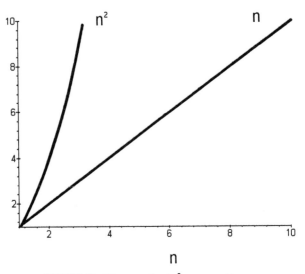

FIGURE 23. The growth of $n^2$ compared to $n$.

$1^2, 2^2, 3^2, 4^2, \ldots$ with the sequence $1, 2, 3, 4, \ldots$, which are simply the numbers in the first sequence unsquared. An easy way to make this comparison is to plot the values of the first sequence on the vertical axis and the second sequence on the horizontal axis. Doing this, we get the graph in Figure 23. Notice that our graph increases at an ever steeper angle. From this we realize that the sequence $1^2$, $2^2, 3^2, 4^2, \ldots$ is increasing much faster than $1, 2, 3, 4, \ldots$. The graph shows us this fact all at once, allowing our minds to grasp it instantly. Using graphs becomes a convenient way to understand how fast sequences and series are increasing.

We now consider the graphs for both $y = e^x$ and $y = \ln x$ ($= \log_e x$). In Figure 24 we have plotted both with the $y$ values on the vertical scale and the $x$ values on the horizontal scale. We have also drawn in the dotted diagonal line $y = x$. Notice that $e^x$ and $\ln x$ are mirror images of each other around that line. The function $e^x$ grows very rapidly and is characteristic of exponential growth. On the other hand, $\ln x$ grows slowly and is characteristic of logarithmic growth. Therefore, exponential growth is rapid, while logarithmic growth is slow.

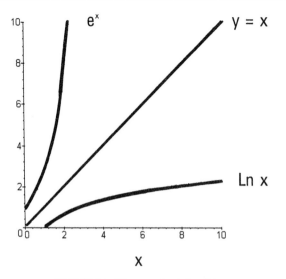

FIGURE 24. The growth of $e^x$ and $ln\ x$.

Now let's graph *log x*, and *ln x* together on the same graph to see how they compare. Remember that *log x* has a base of 10 while the base of *ln x* is *e*. We have plotted the value of both *log x* and *ln x* on the vertical axis and *x* on the horizonal axis to get the graph in Figure 25. We must remember that both *log x* and *ln x* are exponents, i.e., *log x* is the exponent of 10 which yields *x*, or $10^{log\ x} = x$. This means that as *x* grows, log *x* (as an exponent) grows ever slower. This is exactly what happens in Figure 25. In other words, when *x* is 1, we have a vertical value corresponding to zero because $10^0 = 1$. When *x* is 10, then the vertical value will be 1 since $10^1 = 10$. Checking the graph on Figure 25, we do see that the line for *log x* is 0 when *x* is 1 and is 1 when *x* is 10.

When *x* is less than 1, the corresponding values of *log x* and *ln x* are negative. The graph for *ln x* increases faster than *log x* because the base of *e* is smaller than the base of 10. Hence, when *x* = 10 then *ln x* = 2.302585 because $e^{2.302585} = 10$.

Looking at the graphs in Figures 24 and 25 we should be reminded of something we talked about earlier. When we graphed

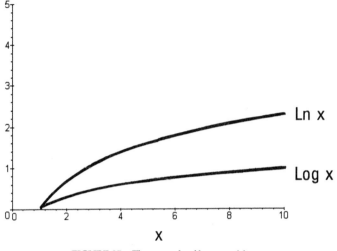

FIGURE 25. The growth of log $x$ and $ln$ $x$.

the increasing values of the harmonic series, we also came up with a graph which increased slowly as we added more and more terms. Figure 26 shows the growth of both the harmonic series, $\Sigma$ $1/n$, and the natural logarithmic function, $ln$ $x$. Notice that the two graphs seem to be increasing at the same general rate and a constant difference. Can we use the logarithmic function to track the growth of the harmonic series? Yes, we can.

Here, again, we have stumbled onto a wonderful connection within mathematics. We started with the harmonic series which grows without limit, but ever so slowly. Now, beginning with $e$, the limit to our interest compounding problem, we have discovered that $e$ (in the form of its logarithm) is connected to the harmonic series. In fact we have the following beautiful limit for $\Sigma$ $1/n$ and $ln$ $n$.

$$\lim_{n \to \infty} \left\{ \sum_{j=1}^{n} \frac{1}{j} - ln\ n \right\} \approx 0.5772157 \ldots$$

Within the parentheses is the difference between the harmonic series and the natural logarithm of $n$. What this limit says is that as

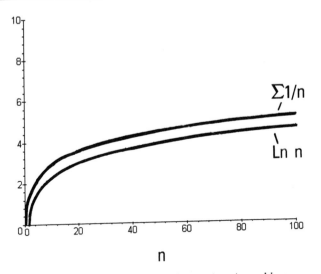

FIGURE 26.   The growth of the harmonic series and $\ln x$.

we take more and more terms to add in the harmonic series, the difference between the harmonic series and the natural logarithm of $n$ approaches a constant. This constant is designated by the Greek letter gamma ($\gamma$). The limit $\gamma$ is known as the Euler–Mascheroni constant and has a value of approximately $0.5772157\ldots$, however, at this time no one knows if $\gamma$ is rational, algebraic, or transcendental.

We now have a method to calculate roughly how large the harmonic series grows for any number of terms. We can use the following relationship:

$$\sum_{j=1}^{n} \frac{1}{j} \approx \ln(n) + 0.5772157$$

where $\sum 1/j$ is the harmonic series after adding $n$ terms. Suppose we want the approximate value of $\sum 1/j$ after adding a million terms? This becomes:

$$\sum \frac{1}{j} \approx \ln(1{,}000{,}000) + 0.5772157$$

or

$$\sum \frac{1}{j} \approx 13.8155105 + 0.5772157 = 14.3927262$$

Suppose we added the first billion terms of the harmonic series, what would we get?

$$\sum \frac{1}{j} \approx ln\ (1,000,000,000) + 0.5772157\ \text{or}$$

$$\sum \frac{1}{j} \approx 20.7232658 + 0.5772157 = 21.3004815$$

After adding a billion terms the sum is still less than 22. Can anyone fail to comprehend how very slowly the harmonic series is diverging? Previously, we defined the function $H(x)$ as the number of terms we must add in the harmonic series to reach a value of $x$. We then went on to ponder the size of $H(\text{googol})$ and $H(\text{googolplex})$ where googol $= 10^{100}$ and googolplex $= 10^{\text{googol}}$. We are now in a position to estimate the size of $H(x)$ for these two values. Using our limit equation we know that:

$$\sum \frac{1}{j} = \text{googol} = 10^{100} \approx ln\ n + 0.5772157$$

All we have to do is solve for $n$, or:

$$ln\ n = 10^{100} - 0.5772157$$

The large value of $n$ needed for the natural logarithm of $n$ to reach $10^{100}$ is so great that we can effectively ignore the Euler-Mascheroni constant. This gives us $ln\ n \approx 10^{100}$. What is $n$? By the definition of logarithms, the above expression is equivalent to:

$$n = e^{10^{100}}$$

For the googolplex we have $ln\ n = 10^{\text{googol}}$ or:

$$n = e^{10^{10^{100}}}$$

Even though the number of terms, $n$, needed to reach a googol and a googolplex are large, what is evident is the fact that we can add enough terms to reach these numbers. This illustrates that the harmonic series really does diverge and can, with enough terms, sum to any number we desire.

We have covered much in this chapter. The main point we want to take away with us is the fundamental nature of $e$, and the natural logarithm function based on $e$. The growth of $e^x$ and $\ln x$ help describe how other mathematical functions behave. In fact, we will come to see that the natural logarithm reaches even more deeply into the workings of mathematics, illustrating even more basic ideas than the harmonic series.

## CHAPTER SIX

# EXOTIC CONNECTIONS

*Natural philosophy, mathematics and astronomy,*

*carry the mind from the country to the creation, and*

*give it a fitness suited to the extent.*

THOMAS PAINE

*ADDRESS TO THE PEOPLE OF ENGLAND*[1]

## THE GOLDEN MEAN

*D*iscovery of the incommensurability of the diagonal of the square sent a shock wave through Greek mathematics, and discredited Pythagorean metaphysics. However, the next logical extension of the same idea gave the Greeks a treasured geometrical concept, and a beautiful number to be handed down through the ages for our enjoyment today. To find this number, we only have to ask: what happens if, instead of a square, we consider a rectangle with sides equal to 1 and 2? Once the Greeks had considered the square with sides equal to one, it is only natural they would have extended this idea to look at the 1×2 rectangle.

In Figure 27 we have such a rectangle, and we have drawn in the rectangle's diagonal. Notice that the diagonal cuts the rectangle into two right triangles. Knowing that the sides of the triangles are 1 and 2, we can use the Pythagorean theorem to compute the length of the diagonal.

$$(\text{diagonal})^2 = 1^2 + 2^2 = 1 + 4 = 5$$

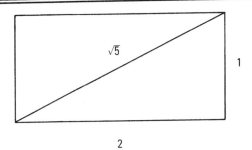

FIGURE 27. Construction of a Golden Rectangle. Begin with a rectangle with sides in the proportion of 1:2, then draw the diagonal whose length is $\sqrt{5}$.

Now we take the square root of each side and get: diagonal = $\sqrt{5}$. But the Greeks didn't stop here. Using one of the triangles, they disconnected one corner, straightened out the side equal to 1, and then rotated the side equal to 2 to form two sides of a new rectangle (Figure 28). The resulting rectangle has a number of interesting features, and, for the ancient Greeks, represented a geometrical shape that was very pleasing to the eye. The proportion or ratio between the two sides of this rectangle is designated by the Greek letter phi ($\phi$). Thus we have:

$$\phi = \frac{\sqrt{5} + 1}{2} \approx 1.6180339\ldots$$

Since the equation for $\phi$ contains a radical, the resulting number is algebraic, but not rational, i.e., it is not equal to the ratio of two whole numbers. Therefore, its decimal expansion, like $\pi$ and $e$, is an infinite, nonrepeating decimal.

The Greeks referred to this ratio by the rather long phrase "the division of a segment in mean and extreme ratio," or sometimes simply called it "the section."[2] If we divide a line segment according to this ratio, we get a rather startling result. The top of Figure 29 shows a line segment $AB$. We divide this segment according to the ratio $\phi$ and mark this division with $C$. If we take the longer segment, $AC$, and divide it by the shorter segment, $CB$, then we get $\phi$, or $AC/CB = \phi$. Now pretend there is a hinge located at point $C$ and rotate the line segment $CB$ around so that it folds back onto

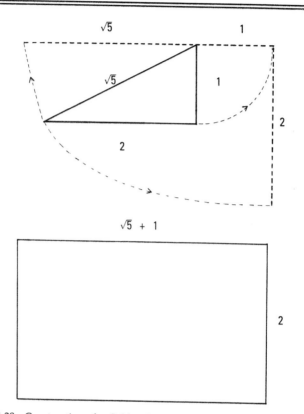

FIGURE 28. Construction of a Golden Rectangle. Disconnect the triangle at the small-angled corner and swing the sides around to form the new rectangle, whose sides are now in the proportion of ($\sqrt{5}$ + 1) to 2.

segment $AC$ as in the middle line of Figure 29. Now the new point $B'$ divides the segment $AC$ into two smaller segments. What is the new ratio of $B'C/AB'$? You guessed it—it's just $\phi$ again.

We can try this folding a second time. On the bottom line of Figure 29 we have folded the line segment, $AB'$, so that new point $A'$ now divides $B'C$. The ratio $B'A'/A'C$ becomes $\phi$ again. And this folding can continue indefinitely. Each time we use the smaller segment to subdivide the larger segment, we get a new subdivision in the same magical ratio of $\phi$.

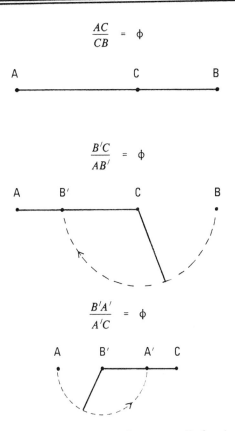

FIGURE 29. Folding the Golden Mean on a line segment. Each swing creates a new line segment whose parts are still in the ratio of φ.

The Greek "section," or ratio of φ, found its way into both Greek architecture and Renaissance art. For example, the ratio of the length to the height of the face of the famous Parthenon in Athens, built in the fifth century B.C., is almost exactly φ. The fact that a rectangle whose sides are in the ratio of φ (which we call a Golden Rectangle) is pleasing to the human eye has been well known for centuries. During the 19th century a number of psychologists, beginning with Adolf Zeising, tested human tastes as they related to the shape of rectangles. Universally they found that we prefer the shape of rectangles close to, or equal to the Golden Rectangle.[3]

Evidence exists that the ratio may have been known to the ancient Egyptians, for the Rhind Papyrus (ca. 1650 B.C.) refers to a "sacred ratio," and the ratio of the altitude of a face of the Great Pyramid at Gizeh to half the length of the base is almost exactly 1.618.[4]

Through the ages other names have been attached to this wonderful ratio including Golden Ratio, Golden Mean, and Divine Proportion. We shall call it the *Golden Mean*.

Now for an intriguing attribute for a rectangle based on the Golden Mean. Look at the rectangle in Figure 30, which has sides equal to 2 and ($\sqrt{5}$ + 1). We have subdivided this rectangle into a square with each side having a length of 2, and a smaller rectangle. The smaller rectangle now has the same ratio to its sides as the larger rectangle, and is, therefore, a smaller Golden Rectangle. Just as we did with the line segment, we can continue this process indefinitely. We subdivide the smaller rectangle into a square, and another, even smaller Golden Rectangle. The process never ends and generates for us an infinite set of ever smaller Golden Rectangles. What is the ratio of the area of one of the Golden Rectangles to the area of the next smaller Golden Rectangle? It is just $\phi^2$.

The Greeks were certainly aware of many of the attributes of $\phi$. The Pythagoreans used a pentagram as one of their holy symbols,

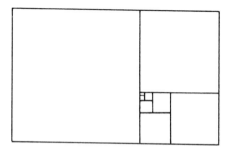

FIGURE 30.  By subtracting out the area of a square from a Golden Rectangle, we form a new, smaller, Golden Rectangle.

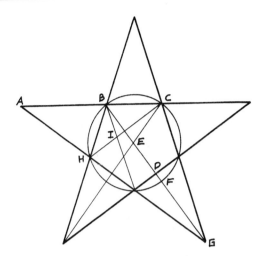

FIGURE 31. The pentagram, used as a symbol by the Pythagoreans, contains many Golden Means. $\phi = AB/BC = CH/BC = IC/HI = 2DE/EF = EG/2DE = \sqrt{E\Gamma/E\Phi}$.

and the pentagram is occasionally used today in the practice of witchcraft and occult ritual. Figure 31 shows a pentagram with different magnitudes identified. The relationships between these magnitudes is most often the Golden Mean.

$$\phi = \frac{AB}{BC} = \frac{CH}{BC} = \frac{IC}{HI} = \frac{2DE}{EF} = \frac{EG}{2DE} = \sqrt{\frac{EG}{EF}}$$

With all these interconnections resulting in a value of $\phi$, is it any wonder the Pythagoreans believed the pentagram was sacred?

Another startling feature of the Golden Mean is that we produce its square by simply adding the number 1. Hence we have:

$$\phi^2 = \phi + 1$$

If we move all terms to the left of the equal sign we get the quadratic equation $\phi^2 - \phi - 1 = 0$. When we solve this equation for $\phi$ using the quadratic formula we get:

$$\phi = (1 + \sqrt{5})/2 \text{ and } \phi = (1 - \sqrt{5})/2$$

Therefore, not only does the Golden Mean satisfy the relation $\phi^2 = \phi + 1$, a second value that produces its own square by adding 1 is the number $(1 - \sqrt{5})/2$ which is approximately equal to minus 0.6180339. . . . This second value, $(1 - \sqrt{5})/2$, is designated as $\phi'$ and is the negative inverse of $\phi$. This means that it satisfies the following equation:

$$\phi \cdot \phi' = -1$$

In addition we have the following nice relationship between $\phi$ and $\phi'$:

$$\phi + \phi' = 1$$

We can generalize the relationship $\phi^2 = \phi + 1$ by multiplying both sides by $\phi$ to get:

$$\phi^3 = \phi^2 + \phi$$

Therefore, to get the cube of the Golden Mean all we have to do is add the square of its value to itself. In fact, we have the rather astounding relationship:

$$\phi^n = \phi^{n-1} + \phi^{n-2}$$

Hence, to compute any power of the Golden Mean we simply add the two immediate lower powers. From this we can generate an entirely new sequence of numbers based on the powers of the Golden Mean, where each term is computed by adding the two previous terms. We begin with the numbers 1 and $\phi$.

Golden Mean sequence = $\{1, \phi, \phi^2, \phi^3, \phi^4, \ldots\}$

Other than the first term, 1, all the succeeding terms are algebraic numbers but not rational numbers. We can actually rewrite this sequence in terms of just $\phi$. We do this by noting:

$$\phi^2 = \phi + 1$$

$$\phi^3 = \phi^2 + \phi = (\phi + 1) + \phi = 2\phi + 1$$

$$\phi^4 = \phi^3 + \phi^2 = (2\phi + 1) + (\phi + 1) = 3\phi + 2$$

$$\phi^5 = \phi^4 + \phi^3 = (3\phi + 2) + (2\phi + 1) = 5\phi + 3$$

Hence, we can write the sequence as just:

$$\{1, \phi, \phi+1, 2\phi+1, 3\phi+2, 5\phi+3, 8\phi+5,...\}$$

We have only scratched the surface of all the different and surprising places where the ratio $\phi$ turns up. However, it is time to move on.

## LEONARDO OF PISA

Ancient Greek mathematics began with the first wise man of Greece, Thales of Miletus (ca. 634–548 B.C.), who established the first Greek school of higher learning on the Mediterranean shores of what is now western Turkey. This may have occurred around 580 B.C. when he was in his 40s. Thus began a long tradition of Greek schools where rich Greek merchants and political leaders sent their children to receive the best education available in the ancient world. In 387 B.C. Plato established his famous Academy at Athens.

After more than a thousand years of Greek achievement and excellence in science, learning, and mathematics, Justinian I, emperor of the Eastern Roman Empire, in 529 A.D. ordered that all pagan philosophical schools be closed. Hence, the famous Academy in Athens was shut down and its property confiscated, ending the great tradition founded 1100 years earlier by Thales. This ushered in the history of medieval Europe, a period lasting approximately 900 years and characterized by a distinct lack of meaningful advancement in mathematics. The work and responsibility of preserving the great achievements of the Greeks and making further contributions to the sciences and mathematics was carried out by the Arabs, who welcomed scholars of all nationalities into their society. Not until the 16th century do we see a general revival of mathematics in Europe. In 1494, Lucas Pacioli, a Tuscan monk, wrote *Summa de Arithmetica*, a compilation of the mathematical knowledge of his day. This ushered in the 16th century and such outstanding mathematicians as Tartaglia, Copernicus, Stifel, Cardan, Recorde, Galileo, and Stevin.

One shining exception during this barren period from 529 until 1500 was the merchant/mathematician Leonardo of Pisa (ca. 1170–1240) who was also known by the name of Fibonacci. Since the

the time of the Romans, Europe had used the Roman numerals and awkward unit fractions for computation. Since Roman numerals do not lend themselves to fast and easy computation, the actual calculations were carried out on counting boards or abacuses, the results then recorded on parchment. While Europeans stumbled around in the dark, the Arabs absorbed into their mathematics the fine Hindu numerals, including zero, which made computation much easier.

Fibonacci's father was a Pisan merchant who was also a customs officer for the North African city of Bugia. Pisa, Venice, and Genoa were the great commercial centers for the Mediterranean during the 12th and 13th centuries, and their merchants enjoyed the freedom to trade throughout the Byzantine Empire. Leonardo took advantage of this freedom and visited many of the area's centers of learning, including Egypt, Greece, Sicily, and Syria. From these visits he learned both the mathematics of the scholars and the calculating schemes in popular, commercial use. In 1202 he published the first of his four mathematics books, *Liber abaci*, which used the Hindu–Arabic numbering system, introducing the Indian numbers 1 through 9 and zero to a European audience. The Muslim influence is clearly demonstrated in his books. The Arabs wrote from right to left. Fibonacci wrote his Indian numerals in descending order, and his mixed fractions with the fraction coming first, i.e., in the form of $\frac{1}{2}4$ instead of the modern custom of $4\frac{1}{2}$.

Although Fibonacci had the foresight to introduce into Europe the superior Hindu–Arabic numerals and do original work in mathematics, his overall impact was less than desired. Europe was slow to adopt the new numerals, and his skill in mathematics was lost to his contemporaries. It would take almost three hundred more years before the Europeans would reach the level of mathematical sophistication necessary to fully appreciate his work. In fact, from 529 when the Academy was closed until 1500 we encounter no significant European mathematician except Fibonacci, though numerous Eastern and Asian mathematicians were practicing, including Brahmagupta from India, Omar Khayyam from Persia, and Tsu Ch'ung-chi from China. At first glance, Europe's hesitancy in accepting Hindu–Arabic numbers appears strange. Yet, there were forces resisting any change. In the traditional system, calculations were made by the mathematician or accountant on an abacus. When the result was known, it could be written on the parchment contract or receipt in Roman numerals. However, when the contract or receipt was reviewed at a later time, only the results of the calculations were preserved, while the method was lost. Since the Hindu–Arabic numerals lent themselves to direct calculation, these calculations could be written upon the parchment to become part of the contract or receipt. This allowed for later checking of calculations to look for mistakes or fraud. Hence, a natural inducement existed to adopt the new numerals. However, the inclusion of calculations on documents significantly increased the amount of parchment consumed, and parchment was expensive. The introduction of cheap paper finally made it possible to record calculations at little additional expense. Not surprisingly, paper was introduced several centuries earlier in the Muslim world than in Western Europe.

However, it is not Fibonacci's use of the Hindu–Arabic numerals, nor his general skill as a mathematician that we want to consider here. Rather, we are more interested in a problem he proposed, and the resulting solution. Within the *Liber abaci* we find the following:

> How many pairs of rabbits will be produced in a year, beginning with a single pair, if in every month each pair bears a new pair which becomes productive from the second month on?[5]

In Figure 32 we have sketched the reproduction of rabbits beginning with our first pair and proceeding through seven generations. We begin with an immature pair and assume that no rabbits die. Each row is one generation, while each column is the life history of a single pair. After a pair is born, it takes another generation for them to mature, after which they produce a new pair each succeeding generation. On the far right of Figure 32 we

○ Non-reproductive Pair
⊗ Reproductive Pair

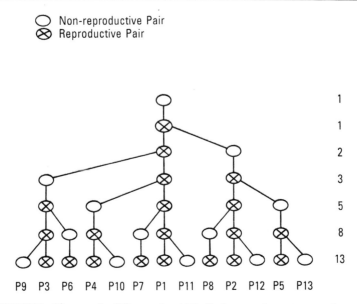

FIGURE 32. The growth of Fibonacci's rabbits. Each new pair must mature for one month before reproducing, after which it will produce one pair each month.

have the number of pairs of rabbits for each generation. What we see at once is that the number of pairs at each generation is the sum of the pairs for the two previous generations. This is always the case, no matter how many generations we wish to extend the chart.

The sequence of numbers we generate in this fashion is:

$$1, 1, 2, 3, 5, 8, 13, 21, 34, 55, 89, 144, 233,\ldots$$

This is called the Fibonacci sequence, and has generated so much interest in the years since Fibonacci first suggested it that a Fibonacci Society was founded in 1962, and a journal, *The Fibonacci Quarterly*,[6] first appeared in 1963, dedicated to unraveling its secrets—and its secrets are many.

Not only does this sequence of numbers have many fascinating mathematical properties, but it seems to be a blueprint used frequently by nature in the growth and generation of living organisms. Some of its mathematical characteristics are quite subtle. For example, any two consecutive Fibonacci numbers will be prime to each other. In other words, no two consecutive Fibonacci numbers will share the same factor. In addition, we know that every prime number will evenly divide an infinite number of Fibonacci numbers. Hence, every prime is contained as a factor in the sequence.

It is customary to designate Fibonacci numbers as $F_n$ where $n$ represents the $n$th term. Hence: $F_1 = 1$, $F_2 = 1$, $F_3 = 2$, $F_4 = 3$, $F_5 = 5$, and $F_6 = 8$. The sum of the first $n$ Fibonacci numbers can be calculated easily:

$$\sum_{i=1}^{n} F_i = F_1 + F_2 + F_3 + \ldots F_n = F_{n+2} - 1$$

To see the beauty of this astounding relationship we look at an example. The Fibonacci number 3 is the fourth term. If we want the sum of all terms through the fourth term, we simply look at the sixth term, the number 8, and subtract 1. Hence, the sum through the first four terms is 7. When we look at any term in the sequence and wish to know the sum of all Fibonacci numbers up to and including that term, we just jump two terms ahead and subtract 1.

This next relationship is no less amazing:

$$\sum_{i=1}^{n} F_{2i} = F_2 + F_4 + F_6 \ldots F_{2n} = F_{2n+1} - 1$$

To add the first $n$ even terms, we just start at term $F_{2n}$ and then jump ahead to the next term and subtract 1. Therefore, if we want the sum of the first five even terms, we look at term 10 (= 55), jump ahead one term to term 11 (= 89) and subtract 1, yielding 88. Therefore, the sum of the first five even terms is 88.

For the odd terms we have an even simpler expression:

$$\sum_{i=1}^{n} F_{2n-1} = F_1 + F_3 + F_5 + \ldots F_{2n-1} = F_{2n}$$

Thus, when we wish to find the sum of the first $n$ odd terms, we go to term $2n$, and that is the answer. There is even a simple relationship for the sum of the consecutive Fibonacci numbers all squared.

$$\sum_{i=1}^{n} F_i^2 = F_1^2 + F_2^2 + F_3^2 + \ldots F_n^2 = F_n F_{n+1}$$

From this, we realize that the Fibonacci sequence is truly a lovely and special sequence. Nature must recognize this fact, for she has incorporated the Fibonacci sequence into the natural world. The sequence is found in everything from the growth of honey bee populations to snail shells, and the arrangement of leaves on plants.[7] Yet, we can go even further. We began this chapter with a discussion of $\phi$, the Golden Mean. Now, why would we then switch to a completely unrelated topic of Fibonacci numbers? Can you see what is coming? The Golden Mean and the Fibonacci sequence are connected in a direct and fundamental way. Let's take consecutive Fibonacci numbers and form their ratios by dividing each one by the previous one.

$F_2/F_1 = 1/1 = 1$
$F_3/F_2 = 2/1 = 2$
$F_4/F_3 = 3/2 = 1/5$
$F_5/F_4 = 5/3 = 1.6666\ldots$
$F_6/F_5 = 8/5 = 1.6$
$F_7/F_6 = 13/8 = 1.625$
$F_8/F_7 = 21/13 = 1.61538$

Can you see where the ratios on the right are going? As we use higher and higher consecutive Fibonacci numbers we get closer to the Golden Mean of approximately $1.61803\ldots$. In fact, the limit of the terms in the right column is exactly the Golden Mean.

$$\lim_{n \to \infty} \frac{F_n}{F_{n-1}} = \phi$$

Looking at the above relationship makes the hair stand up on the back of my neck. How can this relationship be? We start with the special ratio, $\phi$, discovered by the Greeks, and then discover that it is magically related to the Fibonacci sequence. Here, again, we have stumbled onto that very feature of mathematics that makes it so charming and alluring to those willing to make a minimal effort to understand its great secrets.

Remember that we showed how $\phi$ was one of the solutions to the quadratic equation, $X^2 - X - 1 = 0$, and that the other solution was $\phi'$, the negative inverse of $\phi$. We can now use $\phi$ and $\phi'$ to compute the $n$th term of the Fibonacci sequence with the beautiful equation:

$$F_n = \frac{(\phi)^n - (\phi')^n}{\phi - \phi'} = \frac{(\phi)^n - (\phi')^n}{\sqrt{5}}$$

The above equation allows us to compute $F_n$ directly without calculating all the terms less than $F_n$. For example, let's compute $F_{20}$.

$$F_n = \frac{(1.618034)^{20} - (-0.618034)^{20}}{\sqrt{5}} = 6765$$

A practical limit is imposed on the use of this equation to generate the Fibonacci terms since $F_n$ grows quickly in size as $n$ increases. $F_{200}$ is the number:

280571172992510140037611932413038677189525

and $F_{2000}$ is a number 418 digits long!

## THE LUCAS SEQUENCES

The Fibonacci sequence happens to be just one of a whole class of sequences with interesting characteristics, called Lucas sequences. These sequences are named after Edouard Lucas (1842–1891) who studied them extensively. To understand the Lucas sequences we return to the quadratic equation whose solution gives us not only the Golden Mean, $\phi$, but the convergence of the limit of $F_{n+1}/F_n$ as $n$ grows to infinity. This quadratic equation is simply $X^2 - X - 1 = 0$. This is a specific case of a more general equation: $X^2 - PX + Q = 0$, where $P$ and $Q$ are integers not equal to zero. Different pairs of integers, $P$ and $Q$, will define different quadratic equations, giving us an infinite set of solutions. Let $\alpha$ and $\beta$ be the two solutions associated with the integers $P$ and $Q$. We then define the two Lucas sequences $U$ and $V$ in the following way:

$$U_n(P, Q) = \frac{\alpha^n - \beta^n}{\alpha - \beta} \qquad V_n(P, Q) = \alpha^n + \beta^n$$

If we let $P = 1$ and $Q = -1$ then $\alpha = (1 + \sqrt{5})/2$ and $\beta = (1 - \sqrt{5})/2$ which, when substituted into the above equation on the left, yields the Fibonacci sequence. If we use $(1 + \sqrt{5})/2$ and $(1 - \sqrt{5})/2$ in the equation on the right we get the following sequence:

$$2, 1, 3, 4, 7, 11, 18, 29, 47, 76, \ldots$$

Sometimes this sequence is called the Lucas sequence, and the terms are designated as $L_1, L_2, L_3, \ldots$. A casual inspection shows that, it, too, has the property that each term is the sum of the two previous terms or $L_n = L_{n-1} + L_{n-2}$. If we form the fractions in the above Lucas sequence of $L_{n+1}/L_n$ what will they converge to if the Fibonacci terms converge to $\phi$? You guessed it—the Lucas ratios also converge to $\phi$.

It can get even weirder! Take any two positive integers and form the sequence made from adding the two previous terms. This sequence will also have ratios converging to $\phi$. What an astounding number the Golden Mean turns out to be!

## PASCAL'S TRIANGLE

As we have indicated, new ideas in mathematics often arise because someone is trying to solve a particular problem. One such problem is the multiplication of a binomial by itself. Suppose we begin with the simple expression, $x + 1$, where $x$ is some unknown number. Our particular problem requires us to square the expression $x + 1$ which is the same as expanding $(x + 1)^2$. We do this by setting the problem up in the following manner.

$$(x + 1) \cdot (x + 1) = ?$$

Our procedure is to multiply every term in the first set of parentheses by every term in the second set of parentheses and then collect like terms. Doing this we get:

$$(x + 1)\cdot(x + 1) = x^2 + x + x + 1^2 = x^2 + 2x + 1$$

What we want to note here is that the values of the coefficients for the squared terms, i.e., the $x^2$ and $1^2$, are 1 while the coefficient for the $x$ term is 2.

This wasn't really that difficult. But suppose we wanted to take the same expression and cube it. What would we get?

$$(x + 1)^3 = (x + 1)\cdot(x + 1)\cdot(x + 1) = x^3 + 3x^2 + 3x + 1$$

Here the coefficients for the first and last terms are 1 and the coefficients for the two middle terms are both 3. This expansion is harder to do, but not that difficult. Of course, you know we are going to ask: what if we raise $(x + 1)$ to the fourth power?

$$(x + 1)^4 = x^4 + 4x^3 + 6x^2 + 4x + 1$$

We could continue indefinitely expanding to the next higher power, yet things will get very messy, for the number of terms keeps increasing, and soon we are buried in $x$s and coefficients. What we need is to recognize some pattern to the coefficients so that we can easily carry out the expansion of $x + 1$ to any power. This proves useful because certain problems are equivalent to completing such an expansion. At this point we are actually going to go backward and ask what are the coefficients for $(x + 1)$ when the exponents are zero and 1 or: $(x + 1)^0 = 1$, and $(x + 1)^1 = x + 1$. Now we have the coefficients for all expansions from zero to four. We can write just the coefficients for the terms in the following manner.

|  | | | | | | | | |
|---|---|---|---|---|---|---|---|---|
| $(x + 1)^0 =$ | | | | | 1 | | | |
| $(x + 1)^1 =$ | | | | 1 | | 1 | | |
| $(x + 1)^2 =$ | | | 1 | | 2 | | 1 | |
| $(x + 1)^3 =$ | | 1 | | 3 | | 3 | | 1 |
| $(x + 1)^4 =$ | 1 | | 4 | | 6 | | 4 | | 1 |

Do you see any pattern here? Notice that if we take any two adjacent coefficients, their sum is the coefficient between them and one row below.

$$1$$

$$1 \quad + \quad 1$$

$$1 \qquad 2 \qquad 1$$

$$1 \quad + \quad 3 \quad + \quad 3 \qquad 1$$

$$1 \qquad 4 \qquad 6 \qquad 4 \qquad 1$$

Therefore, to get the 2 in the third row we add the two 1s above in the second row. To get the two 3s in fourth row, we add the 1 and 2 above. This pattern continues indefinitely. Armed with this knowledge we can easily write out the expansion of $(x + 1)^5$. The first and last coefficients are always going to be just one. To find those in between, we simply add the two coefficients in the row above. Hence, the expansion of the fifth power of $(x+1)$ becomes: $(x + 1)^5 = 1 \quad 5 \quad 10 \quad 10 \quad 5 \quad 1$. If we enter the appropriate powers for $x$ we get:

$$(x + 1)^5 = x^5 + 5x^4 + 10x^3 + 10x^2 + 5x + 1$$

Of course, if we wanted the expansion using the sixth power we could just write it down by looking at the coefficients for the fifth power. While teaching beginning algebra, I can't restrain myself from having a little fun with the students. I begin expanding the binomial $(x + 1)$, pretending I'm actually figuring out the coefficients in my head. By the time I get to the expansion for the sixth or seventh powers, those students who haven't fallen asleep generally sit up straight in their chairs and shake their heads in wonderment that I can carry out such calculations. Then I begin feeling guilty and confess the secret code that unscrambles everything. I can't do this trick in more advanced classes, because the students recognize this triangle at once, and are on to my shenanigans. The above triangle of coefficients is quite old, and appears to have been discovered independently by both the Persians and the Chinese. The oldest Chinese reference is in the work of Chia Hsien (ca. 1050) which is no longer in existence.[8] Chia Hsien was using the triangle to extract square and cube roots of numbers. The Persian mathematician Omar Khayyam (1048?–11313?), the author of the *Rubaiyat*, probably knew of the triangle since he claimed to have a method for extracting third, fourth, and fifth roots which strongly suggests he was using the triangle. However, the triangle is now known as Pascal's Triangle, named after the French mathematician Blaise Pascal (1623–1662) who made great use of it.

But why are we interested in Pascal's Triangle when we just finished talking about Fibonacci numbers? Could there be a connection between the two? Look at Figure 33. Here is Pascal's Triangle with slanted lines that collect the coefficients into sums that are the Fibonacci numbers. Again we see the magnificent connectedness within mathematics.

We cannot leave Pascal's Triangle without mentioning a connection with prime numbers. Look again at Figure 33. If we go to the third row, we see that the middle term is divisible by 2. In the fourth

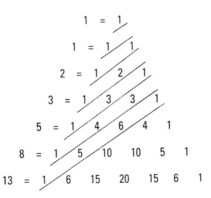

FIGURE 33. The diagonals on Pascal's Triangle add to Fibonacci numbers.

row, the two middle terms are divisible by 3, but in the fifth row, one of the middle terms, the 6, is not divisible by 4. Is this a pattern? Of course it is. In fact we can state: if $n$ is a prime number, then all the middle terms (all terms except the two end terms) of the $n$th +1 row are divisible by $n$. On the other hand, if $n$ is a composite number, then some terms in the $n$th +1 row will not be divisible by $n$. Another mysterious connection.

## CONTINUING FRACTIONS

The Golden Mean and the Fibonacci sequence are fine and good, but we cannot dawdle, we must go on. Now we look at a special way to write fractions that mathematicians use to represent irrational numbers and is also useful in the solution of a special class of polynomials called Diophantine equations, named after the Greek mathematician Diophantus of Alexandria (fl. 250–275). If we begin with the fraction $7/5$ we can rewrite this fraction as $1 + 2/5$. What we have done is to rewrite the original fraction as a whole number plus another fraction, whose numerator and denominator are single digits. Suppose we begin with the number $13/11$? Can we write this as a whole number plus a fraction with a numerator and denominator each less than 10? Let's try.

$$\frac{13}{11} = 1 + \frac{2}{11} = 1 + \frac{1}{\dfrac{11}{2}} = 1 + \frac{1}{5 + \dfrac{1}{2}}$$

The above complex fraction is called a *continued fraction*, because we continue to make fractions in the denominator until we have written our original number in a form that contains only single digits after the first whole number. When we generate a continued fraction where all the numerators are equal to 1, then we have a *simple continued fraction*. The continued fraction for $13/11$ above is such a fraction. Since all numerators are 1, we can write the continued fraction for $13/11$ as a sequence of whole numbers, beginning with the 1:

$$\frac{13}{11} = 1 + \cfrac{1}{5 + \cfrac{1}{2}} = [1; 5, 2]$$

On the right we have [1; 5, 2] which completely specifies how to write the fraction. We separate the first 1 from the rest of the sequence with a semicolon to indicate that it is a whole number while the others represent numbers within denominators of fractions.

Every fraction is equal to a finite simple continued fraction. However, the larger fractions can have continued fractions that are very long and complex. For a more extensive example, let's consider the fraction 237/139. What will its continued fraction look like?

$$\frac{237}{139} = 1 + \cfrac{1}{1 + \cfrac{1}{2 + \cfrac{1}{2 + \cfrac{1}{1 + \cfrac{1}{1 + \cfrac{1}{3 + \cfrac{1}{2}}}}}}} = [1; 1, 2, 2, 1, 1, 3, 2]$$

We can expand the idea of continued fractions by considering infinite continued fractions. These are continued fractions that never end, but continue indefinitely. What kind of number would an infinite continued fraction be? Would it be a number at all? We can actually build the beginning of an infinite continued fraction from a known number. Let's begin with the number $\sqrt{2}$. We can write $\sqrt{2} = 1 + (\sqrt{2} - 1)$. You can see at once that the right hand side of the equation reduces to equal the left side. Now we are going to transform the right side again. We do this by noting the following:

$$(\sqrt{2} - 1) = \frac{(\sqrt{2} - 1) \cdot (\sqrt{2} + 1)}{(\sqrt{2} + 1)}$$

We now multiply the terms together in the numerator and collect like terms to get:

$$\sqrt{2} - 1 = \frac{(\sqrt{2} - 1) \cdot (\sqrt{2} + 1)}{(\sqrt{2} + 1)} = \frac{1}{(\sqrt{2} + 1)}$$

Therefore we can transform $(\sqrt{2} - 1)$ into 1 divided by $(\sqrt{2} + 1)$. Now we return to our original problem where we had:

$$\sqrt{2} = 1 + (\sqrt{2} - 1)$$

We can now substitute into the right side in the following way:

$$\sqrt{2} = 1 + (\sqrt{2} - 1) = 1 + \frac{1}{(\sqrt{2} + 1)}$$

We now rewrite the denominator on the right to get:

$$\sqrt{2} = 1 + (\sqrt{2} - 1) = 1 + \frac{1}{(\sqrt{2} + 1)} = 1 + \frac{1}{2 + (\sqrt{2} - 1)}$$

We again have the term $(\sqrt{2} - 1)$ in the denominator of the far right term, so we can change it into 1 divided by $(\sqrt{2} + 1)$ and begin the process all over again. This will produce the amazingly simple continued fraction:

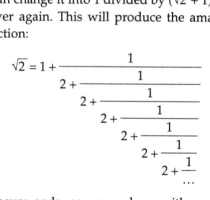

The process never ends, so we end up with a simple infinite continued fraction that is equal to $\sqrt{2}$. With our bracket notation we can write this as $\sqrt{2} = [1; 2, 2, 2, 2, \ldots]$. A customary method to show that a digit (or digits) is repeated indefinitely, as in the case of 2 above, is to place a dot over the repeating digit(s). This gives us the very succinct and beautiful notation of $\sqrt{2} = [1; \dot{2}]$.

And now for the good news! The square root of two is not the only radical with a simple, repeating, infinite continued fraction. Radicals that are square roots are called quadratic surds. One use of the word surd is "devoid of meaning, senseless."[9] Since irrational numbers cannot be represented by a fraction, they were originally considered irrational—not rational or meaningful. Hence, the use of surd to describe them. All simple, repeating, infinite continued fractions are quadratic surds. Therefore, whenever we encounter such a continued fraction, we know that it is equal to some expression containing a quadratic surd. Let's look at some additional examples that are elegant in their simplicity.

$$\sqrt{3} = [1; 1, 2, 1, 2, 1, 2, \ldots] = [1; \dot{1}, \dot{2}]$$

$$\sqrt{5} = [2; 4, 4, 4, 4, \ldots] = [2; \dot{4}]$$

$$\sqrt{7} = [2; 1, 1, 1, 4, 1, 1, 1, 4, \ldots] = [2; \dot{1} \, 1 \, 1 \, \dot{4}]$$

We have discussed how to use continued fractions to represent both fractions (rational numbers) and irrational numbers that are quadratic surds. What about all the other irrational numbers including the strange transcendental numbers? Can they be represented by continued fractions? Yes. Every irrational number, including transcendental numbers, can be represented uniquely by an infinite continued fraction. However, those irrational numbers which are not quadratic surds require a simple, infinite continued fraction which is not periodic, that is, does not repeat itself continuously in a simple pattern.

This is much like the situation we experienced with decimals. Remember that infinite repeating decimals, like 0.3333. . ., were rational numbers (fractions) while nonrepeating, infinite decimals were algebraic numbers, like $\sqrt{2}$, or transcendental numbers, like $\pi$ and $e$. Now we have a *repeating*, infinite continued fraction for certain irrational numbers, quadratic surds like $\sqrt{2}$ and $\sqrt{5}$, but all the rest have nonperiodic, infinite continued fractions.

Even though transcendental numbers do not have simple, periodic, infinite continued fractions, their continued fractions frequently do display a pattern. For example consider the continued fraction for the transcendental number $e$.

$$e = [2; 1, 2, 1, 1, 4, 1, 1, 6, 1, 1, 8, \ldots]$$

In the form of a simple continued fraction, $\pi$ has a very random looking expression. However, if we allow for a continued fraction that is not simple, i.e., one that does not have all ones in the numerators, then we come up with the beautiful expression:

$$\frac{\pi}{4} = \cfrac{1}{1 + \cfrac{1^2}{2 + \cfrac{3^2}{2 + \cfrac{5^2}{2 + \cfrac{7^2}{2 + \cfrac{9^2}{\cdots}}}}}}$$

We can find the limit to a periodic continued fraction as long as that limit is not zero. The procedure is remarkably easy and once you learn it, you can amaze your friends. Suppose we have the continued fraction:

$$L = 4 + \cfrac{5}{4 + \cfrac{5}{4 + \cfrac{5}{4 + \cfrac{5}{4 + \cdots}}}}$$

The first thing we want to do is to divide both sides into 1. Now you see why the limit, $L$, to the continued fraction cannot be zero or we would be dividing by zero!

$$\frac{1}{L} = \cfrac{1}{4 + \cfrac{5}{4 + \cfrac{5}{4 + \cfrac{5}{4 + \cdots}}}}$$

Now we multiply each side by 5:

$$\frac{5}{L} = \cfrac{5}{4 + \cfrac{5}{4 + \cfrac{5}{4 + \cfrac{5}{4 + \cdots}}}}$$

The next step is to add 4 to each side:

$$\frac{1}{5} + 4 = 4 + \cfrac{5}{4 + \cfrac{5}{4 + \cfrac{5}{4 + \cfrac{5}{4 + \ldots}}}}$$

Notice that the right side of the above equation is exactly the same as the right side of the equation we began with. This means we can replace the right side of the equation with $L$, or:

$$\frac{5}{L} + 4 = L$$

After multiplying both sides by $L$ and collecting terms we get the following quadratic equation: $L^2 - 4L - 5 = 0$. Solving for $L$ we find that $L = 2 \pm 3$. Now it makes no sense for $L = -1$, so the answer is $L = 5$. Therefore our continued fraction of fours and fives is equal to 5. This process can be repeated to determine the value of any repeating continued fraction, as long as the limit is not equal to zero.

Of course, we have left the best for last. Once again I remind you that we began this chapter talking about $\phi$ or the Golden Mean. And that is just where we are going to end it. Since the Golden Mean involves a quadratic surd, it must have an infinite continued fraction that is periodic. What could it be? To generate the continued fraction for the Golden Mean all we have to do is remember the relationship $\phi^2 = \phi + 1$. Now we simply divide both sides by $\phi$ to get $\phi = 1 + 1/\phi$. But we can take the entire expression on the right and substitute it into the expression of $\phi$ in the denominator. This yields:

$$\phi = 1 + \frac{1}{\phi} = 1 + \cfrac{1}{1 + \cfrac{1}{\phi}}$$

We can continue to make this substitution indefinitely. Hence, we get:

$$Golden\ Mean = \phi = 1 + \cfrac{1}{1 + \cfrac{1}{1 + \cfrac{1}{1 + \cfrac{1}{1 + \cfrac{1}{\cdots}}}}}$$

That's right! The Golden Mean has the simplest continued fraction of all which can be expressed as simply $\phi = [1; \dot{1}]$.

Finally, we can appreciate how very special the Golden Mean, $\phi$, really is—not only is it a wondrous ratio on its own, but it is also the limit of ratios of successive Fibonacci numbers and the simplest infinite periodic continued fraction. We can write all this symbolically with one more breathtaking equation.

$$\phi = \frac{\sqrt{5}+1}{2} = \lim_{n \to \infty} \frac{F_n}{F_{n-1}} = 1 + \cfrac{1}{1 + \cfrac{1}{1 + \cfrac{1}{1 + \cfrac{1}{\cdots}}}}$$

## ARE RADICALS REALLY IRRATIONAL?

We have had so much fun with the continued fractions, we might as well look at the continued radical. Continued radicals have radicals nested inside other radicals in the following manner:

$$\sqrt{a + \sqrt{b + \sqrt{c + \sqrt{d + \ldots}}}}$$

Notice that we have ended the above formula with the ellipsis, or three dots, to indicate that the process continues forever. Here we go with that strange idea of infinity again! Of course, we will temporarily avoid speaking of infinity by simply asking: Does the above continued form converge to any meaningful value? The simplest case is when the various $a, b, c, d$, etc., are all equal to the same number, $n$. Then the formula becomes:

$$\sqrt{n + \sqrt{n + \sqrt{n + \sqrt{n + \ldots}}}}$$

Does the above formula ever converge? We solve this problem by assuming the formula converges to $L$.

$$L = \sqrt{n + \sqrt{n + \sqrt{n + \sqrt{n + \ldots}}}}$$

Next we add $n$ to both sides.

$$L + n = n + \sqrt{n + \sqrt{n + \sqrt{n + \sqrt{n + \ldots}}}}$$

Now we make a simple but astounding observation. If we take the square root of both sides of the above equation, we get:

$$\sqrt{L + n} = \sqrt{n + \sqrt{n + \sqrt{n + \sqrt{n + \ldots}}}} = L$$

By taking the square root of both sides, the middle of the above expression is just the continued radical we began with, and we said it was equal to the limit $L$. This means that $L$ is equal to the square root of itself plus $n$, or

$$L = \sqrt{L + n}$$

This is beautiful, for we can now square both sides to get:

$$L^2 = L + n$$

Solving for $n$ we finally have: $n = L^2 - L$. Hence, if we want a continued radical whose limit is $L$ we just plug $L$ into the above equation, solve for $n$ and we know our continued radical. For example, suppose we want the continued radical equal to 2. We just plug 2 into the equation:

$$n = 2^2 - 2 = 4 - 2 = 2$$

Hence, $n$ should be 2 or:

$$2 = \sqrt{2 + \sqrt{2 + \sqrt{2 + \sqrt{2 + \ldots}}}}$$

If we want a continued radical for the number 7 we just plug 7 into the equation: $n = 7^2 - 7 = 49 - 7 = 42$. This yields:

$$7 = \sqrt{42 + \sqrt{42 + \sqrt{42 + \sqrt{42 + \ldots}}}}$$

To confirm that this continued radical is actually converging to 7 we compute the first four partial radicals.

$$\sqrt{42} = 6.48074$$

$$\sqrt{42 + \sqrt{42}} = 6.96281$$

$$\sqrt{42 + \sqrt{42 + \sqrt{42}}} = 6.99734$$

$$\sqrt{42 + \sqrt{42 + \sqrt{42 + \sqrt{42}}}} = 6.99981$$

Of course, we don't have to limit ourselves to continued radicals where all the $a$, $b$, $c$, etc., terms under the radical are the same. When we allow the $a$, $b$, and $c$ terms to take on interesting patterns many new and exciting questions and problems spring forth. If we let the terms under the radicals be an increasing sequence, will the continued radical ever converge?

$$\sqrt{1 + \sqrt{2 + \sqrt{3 + \sqrt{4 + \ldots}}}} = ?$$

How about an alternating sequence?

$$\sqrt{1 + \sqrt{2 + \sqrt{1 + \sqrt{2 + \ldots}}}} = ?$$

The possibilities of finding interesting continued radicals are only limited by our imaginations, and we will consider more of them when we look at the work of the Indian mathematician, S. Ramanujan.

We have saved the best continued radical for last. If you have a seatbelt on your seat, now is the time to snap it on tightly, for what you are about to learn has caused many to fall onto the floor in shocked amazement. Since we have a simple relationship, $n = L^2 - L$, that tells us what $n$ to use in the continued radical to achieve a specific limit, $L$, there is nothing stopping us from asking what should $n$ be if we want our limit to be the Golden Mean or $\phi$. Thus we get:

$$n = \phi^2 - \phi$$

Remembering our earlier work with $\phi$ we know that $\phi^2 = \phi + 1$, hence we can substitute the righthand expression in for $\phi^2$ to get:

$$n = (\phi + 1) - \phi = \phi - \phi + 1 = 1$$

Therefore, the correct $n$ that gives us a limit of $\phi$ is just 1.

$$\phi = \sqrt{1 + \sqrt{1 + \sqrt{1 + \sqrt{1 + \ldots}}}}$$

This looks suspiciously like the infinite continued fraction we obtained for $\phi$.

$$\phi = \sqrt{1 + \sqrt{1 + \sqrt{1 + \sqrt{1 + \ldots}}}} = 1 + \cfrac{1}{1 + \cfrac{1}{1 + \cfrac{1}{1 + \cfrac{1}{1 + \ldots}}}}$$

Again we are struck by the role played by the Golden Mean. It is not only the simplest continued fraction, but also the simplest continued radical, both containing nothing but ones. It is hard to deny that something magical surrounds the Golden Mean. As wonderful as the above relationship is between the continued fraction and continued radical for $\phi$, we still have one more result to tickle your mathematical funny bone. Let's see what happens when we find the appropriate $n$ for the limit of 1. That is:

$$n = 1^2 - 1 = 0$$

This says that zero is the correct $n$ for a continued radical equal to 1 or:

$$1 = \sqrt{0 + \sqrt{0 + \sqrt{0 + \sqrt{0 + \ldots}}}}$$

But, you object, the square root of zero is just zero. How can we get 1 from taking all those square roots of nothing? Your objection is well grounded, for we can construct the following infinite sequence:

$$\sqrt{0}, \sqrt{0 + \sqrt{0}}, \sqrt{0 + \sqrt{0 + \sqrt{0}}}, \sqrt{0 + \sqrt{0 + \sqrt{0 + \sqrt{0}}}}, \ldots$$

In this sequence, each finite nested radical can be evaluated as equal to zero. Therefore, the infinite sequence of such zeros has a limit that is zero. This would seem to demonstrate that our finite nested radical of zeros is equal to zero and not 1.

It is the power of mathematics to show us truths that our intuitive minds are slow to grasp. This is just a case. Again consider the general equation we used to determine what $n$ to use in an infinite radical to get a specific $L$ or, $n = L^2 - L$. If we solve this equation for $L$ in terms of $n$ we get:

$$L = \frac{1 \pm \sqrt{4n + 1}}{2}$$

Notice that the above equation yields two answers. If we let $n$ be zero then we get both $L = (1 + \sqrt{1})/2 = 1$ and $L = (1 - \sqrt{1})/2 = 0$. Hence, the algebraic solution tells us there are two different limits to the infinite nested radical containing only zeros. What's going on here? The answer is, it depends on how you construct your radical. If you use an infinite sequence of (finite) nested zeros, you get zero. If, however, you begin with an $n$ in an infinite radical that is something larger than zero, and allow it to decrease to zero, you get the infinite nested radical that has 1 as its limit. We can show this as:

$$\lim_{n \to 0} \sqrt{n + \sqrt{n + \sqrt{n + \sqrt{n + \ldots}}}} = 1$$

While it is true that the square root of zero is zero, it is also true that we can build infinite continued fractions of zeros equal to 1 or zero.

Your may now unsnap your seat belt.

## CHAPTER SEVEN

# CLOSING IN ON THE PRIMES

*You know of course that a mathematical line, a line of*

*thickness nil, has no real existence. They taught you*

*that? Neither has a mathematical plane. These things*

*are mere abstractions.—The Time Traveller*

H.G. WELLS

*THE TIME MACHINE*[1]

To fully comprehend the natural number sequence, we must look once again at prime numbers. We have already mentioned that the Greeks knew the distinction between prime and composite numbers, and even proved useful theorems concerning them. All natural numbers (excluding 1) can be categorized as either prime numbers (primes) or composite numbers (composites). Why are we interested in prime and composite numbers? Because every composite number "decomposes" into a unique set of prime numbers multiplied together. As mentioned earlier, this is the *Fundamental Theorem of Arithmetic*: Every natural number greater than 1 can be expressed as a product of prime numbers in one and only one way.

This is an important theorem for it guarantees that if a number factors into a set of primes, these primes uniquely describe that number. Hence, 2·3 represents only 6. We might write it as 2·3 or 3·2 for the order is unimportant. Therefore, every natural number can be described by its unique set of primes. The unique set of primes that a number factors into are called that number's *prime factors.* Numbers that factor into only one prime (themselves) are, of course, prime numbers. The Fundamental Theorem of Arithmetic is one of the pillars of mathematics, and is used in numerous proofs of other mathematical theorems. Without this theorem, the very laws of algebra would crumble.

## THE FIRST PRIMES

Let's start by looking at the first natural numbers and ask the simple question: Are they prime or composite? We do not consider 1 a prime number even though it technically satisfies the definition of a prime, i.e., being evenly divisible only by 1 and itself. If we were to let 1 be a prime, then the Fundamental Theorem of Arithmetic would break down because there would be different ways to factor the same number. For example we could factor the number 12 as:

$$12 = 2 \cdot 2 \cdot 3$$

$$12 = 1 \cdot 2 \cdot 2 \cdot 3$$

$$12 = 1 \cdot 1 \cdot 2 \cdot 2 \cdot 3$$

We can see what a can of worms is opened by letting 1 be a prime. Such a move allows every number to be factored in an infinite number of ways. Therefore, we place 1 in a special category all by itself. We will simply designate it as unity.

Figure 34 shows the first 100 numbers with the primes circled. We notice some things about primes at once. Of all the primes, only the first prime, 2, is even. All others are odd. We realize that all primes other than 2 must be odd, for if another even prime existed, it would be evenly divisible by 2, and hence, not prime. Of the 15 numbers between 2 and 16, six are primes and nine are composites. Notice that the first numbers are rich in primes and, as the natural numbers get larger, fewer primes are found. This leads to several interesting questions. Do the prime numbers continue to thin out as the natural numbers get bigger? How many prime numbers are there? Do the prime numbers finally disappear entirely? An infinite number of natural numbers exist, for we can, theoretically, count on forever. Do an infinite number of primes exist or only some finite

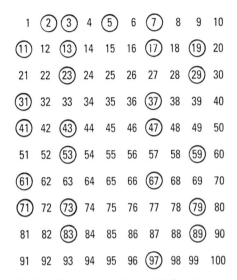

FIGURE 34. The first 100 natural numbers with the prime numbers circled.

number? How are the prime numbers distributed throughout the natural numbers? Is there a predictable pattern to them? Can we formulate a rule or an equation so that we can always calculate exactly what the $n$th prime is for any number $n$?

We can tell whether small numbers are prime or composite simply by inspection. But what about larger numbers? Is the number 8831 a prime or composite number? (It's prime.) What about the number 7,317,943,311? (It's composite.) What we need is a useful procedure to tell whether a number is prime or not. If composite, what are its prime factors? This last question is of special importance to modern code users.

## EUCLID'S PROOF

The answer to the question about the number of primes has been known since ancient times and represents one of the shining gems of Greek mathematics. The great mathematician Euclid proved over 2000 years ago that infinitely many primes exist. His proof was simple and elegant. He began by assuming that only a finite number of primes exist, and then showed that this assumption leads to a contradiction. If only a finite number of primes exist, then one of them must be a largest prime. We will call it $P$. He then constructed another number in the following way. He multiplied all the prime numbers from 2 up to and including $P$ together and then added the number 1. Euclid's number is:

$$2 \cdot 3 \cdot 5 \cdot 7 \ldots \cdot P + 1$$

What kind of number is Euclid's number? If $P$ is assumed to be the biggest prime, then Euclid's number must be composite since it is clearly larger than $P$. If his number is composite then it will be evenly divisible by at least one of the existing primes. Yet, every prime from 2 through $P$ leaves a remainder of 1 when divided into Euclid's number. Therefore, if it really is composite, then some prime larger than $P$ must divide it. Yet, this contradicts the assumption that $P$ is the largest prime. Since the assumption that there exists a largest prime is false, there must be infinitely many primes.

We can also answer the question about primes thinning out. Yes, we do find fewer and fewer primes as we go higher into the natural numbers. Of the first 100 numbers, 25 are prime. This is 25 percent. Yet, in the 100-number gap between the numbers 50,000 and 50,100 there are only ten primes for a prime rate of 10 percent. The percentage of primes does not diminish to some fixed number larger than zero, but continues to decrease and, on average, approaches zero. In mathematical jargon, zero is the limit of this percentage as we proceed up through larger and larger natural numbers. Table 5 shows the percentage of primes for various ranges of natural numbers.

As we proceed higher into the natural numbers and the primes thin out, the strings of consecutive composite numbers between primes grow ever longer. The smallest gap between primes, of course, is between the primes 2 and 3. These primes differ by only one and have no composite between them. This is the only place in the natural numbers where this happens. Between all other consecutive primes there exists a minimum difference of two (such as 5 and 7, and 11 and 13) with at least one composite between the two primes. The difference between consecutive primes (excepting the

**Table 5.** Percent of Numbers That Are Primes

| Range (up to) | Percent Primes | One Prime Out Of | Comments |
|---|---|---|---|
| $10^3$ | 16.80 | 6 | Thousand |
| $10^6$ | 7.85 | 13 | Million |
| $10^9$ | 5.08 | 20 | Billion |
| $10^{12}$ | 3.76 | 27 | Trillion |
| $10^{100}$ | 0.436 | 229 | Factorization Programs |
| $10^{155}$ | 0.281 | 356 | Public-Key Code Numbers |
| $10^{1,000}$ | 0.0434% | 2,300 | |
| $10^{227,832}$ | 0.00019% | 525,000 | Largest Prime Number |
| $10^{1,000,000}$ | 0.0000434% | 2,300,000 | |

primes 2 and 3) is always even, which means the number of composite numbers in the gap between consecutive primes will always be odd.

Between 1 and 100 the largest gap is between the primes 89 and 97 and consists of seven composite numbers. For the numbers less than 1000 the largest gap is between 887 and 907 with 19 composite numbers.

How big do these gaps between primes grow? We can, in fact, find a gap between consecutive prime numbers as large as we want. Given any number, $N$, we can construct a sequence of consecutive composite numbers, at least $N$ numbers long. The procedure is remarkably simple. Let's begin by looking at the following two equations:

$$1 \cdot 2 \cdot 3 + 2 = 8$$

and

$$1 \cdot 2 \cdot 3 + 3 = 9$$

The left sides of the above expressions can be factored into the following forms:

$$2(1 \cdot 3 + 1) = 8$$

and

$$3(1 \cdot 2 + 1) = 9$$

Since the left side of each of the above equations is the product of two numbers, the resulting number is composite. Hence, both 8 and 9 are composite because they can be represented as the product of two natural numbers (excluding 1). We found a sequence of two consecutive composite numbers with our two special number forms, $1 \cdot 2 \cdot 3 + 2$ and $1 \cdot 2 \cdot 3 + 3$. Notice that the 2 and the 3 that we are adding both occur as factors in the product $1 \cdot 2 \cdot 3$. Since $1 \cdot 2 \cdot 3$ can be written as 3! (using factorial notation) we can rewrite the two consecutive composite numbers as $3! + 2$ and $3! + 3$.

If we want a sequence of three consecutive composite numbers we use the three forms $4! + 2$, $4! + 3$, and $4! + 4$. This yields the following:

$$4! + 2 = 1 \cdot 2 \cdot 3 \cdot 4 + 2 = 24 + 2 = 26$$

$$4! + 3 = 1 \cdot 2 \cdot 3 \cdot 4 + 3 = 24 + 3 = 27$$

$$4! + 4 = 1 \cdot 2 \cdot 3 \cdot 4 + 4 = 24 + 4 = 28$$

The three numbers generated (26, 27, and 28) are all consecutive composite numbers. Yet, they are not the first three consecutive composite numbers 8, 9, and 10. This procedure will always give us a sequence of the desired length, but it doesn't produce the smallest such sequence.

If we want to generate a sequence of composites that is $N$ numbers long then we simply write the following forms:

$$(N+1)! + 2$$

$$(N+1)! + 3$$

$$(N+1)! + 4$$

$$\vdots$$

$$(N+1)! + (N+1)$$

The resulting numbers will be $N$ consecutive composites. Unfortunately, this procedure is not very efficient. If we want a long sequence, the resulting numbers in the sequence will be very large. For example, to generate a sequence of ten consecutive composites requires computing 11! which is equal to 39,916,800. By inspection we found a larger sequence of 19 composites between 887 and 907.

If we want to find a sequence of consecutive composite numbers a million numbers long, we can do it. The numbers of the resulting sequence may be extremely large, but that does not detract from the fact that such sequences exist. With such large gaps between prime numbers, we realize at once how sparse prime numbers become when we get into very large natural numbers.

What is the largest gap between primes that has been located through inspection (rather than using the above procedure)? A gap of 803 composite numbers exists between the primes 90,874,329,411,493 and 90,874,329,412,297 which was found in 1989 by J. Young and A. Potler.[2]

## TESTING NUMBERS FOR PRIMENESS

What about testing a number to see if it is prime? A theorem exists for testing numbers for primeness, discovered by Edward Waring but named for his friend, John Wilson. Edward Waring (1734–1798) was a mathematics professor at Cambridge, England, and wrote a famous text, *Meditations Algebraicae*, which contained much material on primes including the theorem he named after John Wilson. Wilson (1741–1793) was also trained as a mathematician at Cambridge but left math for a career in law. He subsequently became a judge and was knighted.

> Wilson's Theorem: $P$ is a prime number if, and only if, $(P - 1)! + 1$ is divisible by $P$.

Here, again, we encounter the factorial symbol in an equation regarding primes. We know that $(P - 1)!$ results from multiplying all the numbers from 1 through $P - 1$ together. Wilson's Theorem states that if we want to test whether a number $P$ is prime, all we have to do is multiply all the numbers less than $P$ together, add 1, and then try to divide by $P$. Let's test it with the prime 5. First we multiply all the numbers less than 5 together: $1 \cdot 2 \cdot 3 \cdot 4 = 24$. Now we just add 1: $24 + 1 = 25$. Will 5 divide 25? Yes, therefore 5 must be prime.

Of course we knew all along that 5 was prime. While Wilson's Theorem will test any number for primeness, there is a serious drawback. We generally want to test very large numbers for primeness. The difficulty of multiplying all the numbers that are less than a very large number to conduct the test makes the theorem impractical. Theoretically, it's a fine theorem, but for us it's of no real use. Therefore, a meaningful question in the study of prime numbers becomes: Are there other procedures that we can use to test a number for primeness which are quicker than Wilson's Theorem? Fortunately, the answer is yes.

Can we tell if a number is composite just by looking at it? In many cases we recognize a composite by simply looking at its far right digit. Two is the only even prime number. Hence, if a number ends on the right with a 0, 2, 4, 6, or 8 we know it is composite since all such numbers are even. Therefore we can immediately eliminate 1/2 of all natural numbers from consideration as primes. Yet, we can do even better than this. All numbers that end with a 5 are divisible by 5. Therefore, once we get beyond the first primes of 2, 3, and 5, we can say that only numbers ending in 1, 3, 7, and 9 may be prime numbers. This reduces to 40% the numbers we must test for primeness.

One way to test if a number, $N$, is prime is to try and divide $N$ by all those numbers (excluding 1) less than $N$. If one of them evenly divides $N$, then $N$ must be composite. Suppose we wish to know if the number 839 is a prime? Using this method we try to divide 839 by all the numbers from 2 up to and including 838. If none of them divide evenly, then 839 must be prime. This procedure involves 837 attempted divisions. Therefore, testing a number $N$ for primeness with this method commits us to as many as $N - 2$ operations. Of course, if $N$ turns out to be composite, then at some point one of our divisions will come out even and we can stop. This procedure is certainly easier than the one required by Wilson's Theorem, yet it is still too long. To test whether 1,000,003 is prime requires a million and one attempted divisions because it is, in fact, a prime number. Can we make improvements on this method?

We have a very nice theorem which is going to cut our work considerably. It's known as the Sieve of Eratosthenes. Eratosthenes (276–195 B.C.) was a Greek astronomer, geographer, and mathematician. Like Euclid, he studied at Athens and then became head of the Alexandrian Library in the Egyptian city of Alexandria. In addition to his Sieve Theorem, which has been used for centuries in testing for primeness, he is best remembered for his remarkable measure of the Earth's circumference, coming very close to the correct figure.

> Sieve of Eratosthenes: If $N$ is a composite number then at least one of the prime factors of $N$ is less than or equal to the square root of $N$.

This theorem says that for every composite number, at least one of the prime factors must be equal to, or smaller than, its square root. It is easy to see why this theorem is true. If a number, $N$, is composed of only two primes, then the largest those two primes could be is when they are the same size and $N$ is a perfect square. Hence, 9 factors into 3·3. Here, the two primes that make up 9 are the same size and both are equal to the square root of 9. If the two primes making up $N$ are different in size, then the smaller has to be less than the square root of $N$, while the other is larger. When we go to numbers with three factors we get an even stronger statement. If we have a number with exactly three prime factors, then the largest these primes can be is the cube root of $N$ when $N$ is a perfect cube. The number 27 has three factors, 3·3·3. Each is exactly equal to the cube root of 27. Any number with three factors that is not a perfect cube, must have one factor less than the cube root of $N$.

This yields a much stronger theorem: If $N$ has $n$ prime factors, then at least one prime factor must be less than or equal to the $n$th root of $N$. The only problem in using this stronger theorem is that when we have a big number and don't know if it is prime or not, then we don't know how many factors it has. Hence, we assume it could have just two factors and use the Sieve of Eratosthenes.

To find a prime factor of a composite, $N$, just test all the numbers from 2 up to the square root of $N$, a substantial decrease in the number of divisions necessary to test for primeness. Consider the number 101: Is it prime? If it is not prime then one of its prime factors must be smaller than or equal to the square root of 101. What are these possible numbers? The numbers less than or equal to

$\sqrt{101}$ are 2, 3, 4, 5, 6, 7, 8, 9, and 10. Hence, we only have to test nine numbers to see if they divide into 101.

We can do even better. The Sieve of Eratosthenes says that the factor less than the square root of $N$ is a prime factor (number). Those prime numbers less than or equal to $\sqrt{101}$ are 2, 3, 5, and 7. We need only try dividing 101 by these four primes to see if it is prime. Carrying out the division we find none of the four primes (2, 3, 5, 7) divide 101 evenly and therefore 101 is, indeed, a prime.

To sum up our testing procedures: We only need test numbers ending in 1, 3, 7, and 9 for primeness. We only have to divide such numbers by those primes less than or equal to the square root of the number being tested. In addition to the above tests, there are a few additional procedures which are of marginal value. For example, if the sum of the digits of a number is divisible by 3 then the original number is divisible by 3. Consider the number 104,001. We can see at once that the digits in this number add to 6 and 6 is divisible by 3. Hence, 104,001 must be divisible by 3. (It is since 104,001 = 3·34,667.) This test is of limited value because we can divide a number by 3 almost as fast as we can add its digits. We have a more complex test for divisibility by 11, which is to alternately add and subtract the digits of a number, moving from left to right. If the result is zero or is divisible by 11, the original number is divisible by 11. We'll try it with 16,401. First we add 1 (the first digit on the left) and subtract 6. This yields a –5. Then we add 4 to get –1. We subtract 0 and add 1 to get the final result of zero. Therefore, the original number (16,401) is divisible by 11 and 16,401 = 11·1491. All this information shortens our testing considerably, yet with large numbers we are still in a pickle, for the time necessary to do the testing can be beyond even the capabilities of modern computers.

## HOW MANY PRIMES?

We are now ready to consider the density of primes within the natural numbers. That is, as the natural numbers grow ever larger, how fast does the number of primes grow? Our main focus will be to arrive at some understanding of how the primes thin out as the

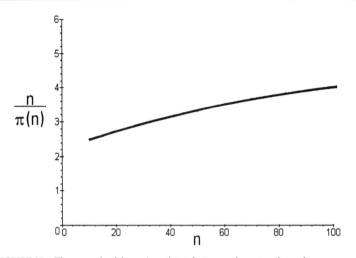

FIGURE 35. The growth of the ratio, $n/\pi(n)$, between the natural number sequence, $n$, and the prime counting function, $\pi(n)$.

natural numbers get larger. We used graphs earlier to show the growth of mathematical functions, and we can do the same here. In Figure 35 we show the natural numbers on the horizontal scale, and the ratio of the natural numbers to the number of primes on the vertical scale. Notice how the graph in Figure 35 bends toward the right. Does this remind us of anything? Of course! When we graphed the harmonic series, we saw the same kind of growth. To measure that growth we discovered we could use the natural logarithmic function. Can we do so here? Will this show another remarkable connection within mathematics, i.e., the growth of prime numbers and the wonderful number $e$?

To understand the density of primes, we must define a new symbol that stands for the number of primes less than some number, $n$. Mathematicians have given this number the name $\pi(n)$, or the *Prime Counting Function*. Do not confuse our Prime Counting Function, $\pi(n)$, with the ratio of the circumference of a circle with its diameter which is designated as simply $\pi$. They are two distinctly different animals, even though they both use the Greek letter pi.

*Prime Counting Function:* π(n), is the number of primes less than or equal to *n*.

Now for the bad news! No straightforward formula exists that will compute π(n) exactly. This has great significance for our understanding of the natural numbers. It means that the specific locations of the primes embedded within the natural numbers is so random that we cannot predict precisely where they will be. As counting animals, we are not used to this vagueness. We know, for example, that beginning with 2 every other number will always be even. We also know, based on our number system, that beginning with 5, every fifth number will be divisible by 5. With great regularity we can predict where digits will fall within numbers because our number system is periodic. Yet, when we try to get a feeling for where the primes are, their strange unpredictability defeats us. Yet, primes are so terribly important because they are the building blocks of all natural numbers.

We have to settle for formulas that give us only approximations for π(n). One such formula that approximates π(n) for very large numbers is a cornerstone to modern number theory and is known as the Prime Number Theorem.

*Prime Number Theorem:* The number of primes less than *n* is approximately *n* divided by the logarithm of *n*.

We can state this in a more symbolic manner with:

$$\pi(n) \approx n/ln(n)$$

In the above equation *ln* of course stands for the natural logarithm with base *e* and not the common logarithm of base 10. The two horizontal wiggly lines stand for "approximately" and are distinct from the more common symbol, the equal sign (=).

For example, if we wish to know the approximate number of prime numbers less than 1,000,000 we compute (1,000,000)/*ln*(1,000,000) which is 72,382. There are exactly 78,498 such primes, so we can see that 72,382 is only a rough approximation, in error 7.8%. However, as we consider larger numbers our function *n/ln(n)* gives us even better approximations. In the Prime Number Theorem we have a powerful function that approximates the growth of prime numbers while at the same time connecting this growth to the natural logarithms. In turn, the natural logarithms are based on the number $e$, itself the limit to the function $(1 + 1/n)^n$, and this function is linked to the growth of the harmonic series. We must begin to wonder: Is there some magical connection between everything mathematical? Carl Gauss conjectured the Prime Number Theorem in 1792 when he was only 15 years old. How did he do it? He counted primes and calculated $\pi(n)$ for consecutive blocks of 1000 numbers and noticed how $\pi(n)$ was diminishing. He was unable to prove the theorem, which was not accomplished until 1896 when two mathematicians, Jacques Hadamard and C. J. de la Vallée-Poussin, independently proved it.

How well does $n/ln(n)$ approximate $\pi(n)$? Table 6 shows both $\pi(n)$ and $n/ln(n)$ for increasing values of $n$ plus the difference and the percent error between $\pi(n)$ and $n/ln(n)$. From this table we see that the absolute difference between $\pi(n)$ and $n/ln(n)$ grows in size as $n$ increases while the percent error decreases. Unfortunately, the percent error decreases slowly, so that even at moderately large $n$ the error for $n/ln(n)$ is still substantial.

**Table 6.** Comparison of $\pi(n)$ and $n/ln(n)$

| $n$ | $\pi(n)$ | $n/\log(n)$ | Difference | Percent Error |
|---|---|---|---|---|
| $10^2$ | 25 | 21 | 4 | 16.00 |
| $10^3$ | 168 | 144 | 24 | 14.29 |
| $10^4$ | 1,229 | 1,085 | 144 | 11.72 |
| $10^5$ | 9,593 | 8,685 | 908 | 9.47 |
| $10^6$ | 78,499 | 72,382 | 6,117 | 7.79 |
| $10^7$ | 664,579 | 620,420 | 44,159 | 6.64 |
| $10^8$ | 5,761,455 | 5,428,680 | 332,775 | 5.78 |
| $10^9$ | 50,847,534 | 48,254,945 | 2,592,589 | 5.10 |
| $10^{10}$ | 455,052,512 | 434,294,493 | 20,758,019 | 4.56 |
| $10^{11}$ | 4,118,054,813 | 3,948,131,889 | 169,922,924 | 4.13 |
| $10^{12}$ | 37,607,912,018 | 36,191,208,672 | 1,416,703,346 | 3.77 |
| $10^{13}$ | 346,065,536,839 | 334,072,662,679 | 11,992,874,160 | 3.47 |
| $10^{14}$ | 3,204,941,750,802 | 3,102,103,502,550 | 102,838,248,252 | 3.21 |
| $10^{15}$ | 29,844,570,422,669 | 28,952,965,081,228 | 891,605,341,441 | 2.99 |

Fortunately, another more accurate formula exists, called the *Logarithmic Integral of n*, designated as $li(n)$. It was also discovered by Carl Gauss, and has the following rather daunting formula:

$$li(n) = \int_{2}^{n} \frac{du}{ln(u)}$$

The above formula is called a *definite integral* from calculus and involves mathematics beyond our consideration here. Don't worry about it, however. There is a convenient way to approximate the value of $li(n)$ without resorting to calculus with the following series:

$$li(n) \approx n \left( \frac{1}{ln\ n} + \frac{1!}{(ln\ n)^2} + \frac{2!}{(ln\ n)^3} + \ldots + \frac{(k-1)!}{(ln\ n)^k} \right)$$

In the above formula, $k$ is the number of terms we add together in the series to approximate $li(n)$. This formula is quite accurate when $n$ and $k$ are fairly large. While it is somewhat messy, it is easily programmed for a computer. In Table 7, $li(n)$ has been approximated using the above formula taken to 12 terms ($k = 12$), and

**Table 7.** Comparison of $\pi(n)$ and $li(n)$

| $n$ | $\pi(n)$ | $li(n)$ | Difference | Percent Error |
|---|---|---|---|---|
| $10^2$ | 25 | 111 | 86 | 344% |
| $10^3$ | 168 | 187 | 19 | 11.309520 |
| $10^4$ | 1,229 | 1,249 | 20 | 1.627339 |
| $10^5$ | 9,593 | 9,630 | 37 | 0.385698 |
| $10^6$ | 78,499 | 78,626 | 127 | 0.161785 |
| $10^7$ | 664,579 | 664,915 | 336 | 0.050558 |
| $10^8$ | 5,761,455 | 5,762,203 | 748 | 0.012982 |
| $10^9$ | 50,847,534 | 50,849,225 | 1,691 | 0.003325 |
| $10^{10}$ | 455,052,512 | 455,055,600 | 3,088 | 0.000678 |
| $10^{11}$ | 4,118,054,813 | 4,118,066,574 | 11,761 | 0.000285 |
| $10^{12}$ | 37,607,912,018 | 37,607,953,542 | 41,524 | 0.000110 |
| $10^{13}$ | 346,065,536,839 | 346,065,632,227 | 95,388 | 0.000027 |
| $10^{14}$ | 3,204,941,750,802 | 3,204,942,067,508 | 316,706 | 0.000010 |
| $10^{15}$ | 29,844,570,422,669 | 29,844,571,135,055 | 712,386 | 0.000002 |

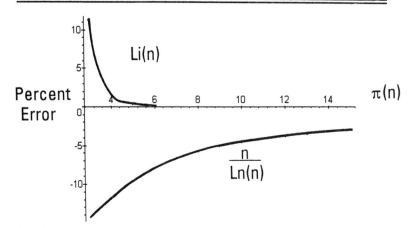

FIGURE 36. The percent error between $li(n)$ and $\pi(n)$; and between $n/ln(n)$ and $\pi(n)$, as $n$ increases. The horizontal scale is in powers of ten and ranges from $10^1$ to $10^{17}$.

illustrates how much better $li(n)$ is at approximating $\pi(n)$ than $n/ln(n)$.

Notice in Table 7 that for small values of $n$, e.g., 100 and 1000, the error in $li(n)$ is large, but that the percent error diminishes to almost nothing as $n$ increases in size. By the time we get to 1,000,000, the error in $li(n)$ is less that 2/10 of 1%. Even though the absolute difference between $\pi(n)$ and $li(n)$ continues to increase, the percent error becomes negligible. Figure 36 demonstrates how $n/ln(n)$ and $li(n)$ approximate $\pi(n)$ as $n$ increases.

Using both $n/ln(n)$ and $li(n)$ to give us estimates of $\pi(n)$ we can try to get a feel for how the primes decrease as the natural numbers increase. Table 5 shows the actual or estimated $\pi(n)$ for blocks of numbers beginning at 1000 and increasing to 10 raised to the millionth power, or the number you get when you write a 1 followed by one million zeros. At the range of 1000 approximately 16% of the numbers are primes. This is roughly one out of six numbers. We have included comments to help us get a feel for just how large the various ranges are.

Up to 1000 is the range in which normal individuals carry on their mathematics, e.g., balancing checkbooks, doing taxes, and estimating mileage. At the next range of one million we find governments computing taxes and buying bombers. Here, only 7.2% of the numbers are prime. The next range is at one billion. Normal people don't use numbers this large, but our government needs such numbers to compute the national budget.

The next range is at one million million or one trillion. Here we end everyday number usage since the only reference is the national debt. One trillion is a one followed by 12 zeros or $10^{12}$. As a comparison, astronomers estimate that there are $10^{11}$ stars in the Milky Way Galaxy and in excess of $10^{18}$ stars in the visible universe.

The next range, one followed by 100 zeros, is the size of numbers that mathematicians can factor into primes with any degree of certainty. This ability has only been achieved in the last few years. At this range only 1/2 of one percent of the numbers are prime or about 1 in 233. For a physical reference, we have the much smaller number, $10^{80}$, which is the estimated number of protons in the universe.

The next range is $10^{155}$. This is the size of numbers used in modern public-key codes, codes used to transmit information in a scrambled format to prevent eavesdropping. At this range, only one number in 357 is prime.

Now we are going to get very big. The next range is at $10^{1000}$. Here we have no useful references. The percentage of primes has dropped to 4/100 of one percent or approximately one out of 2300 numbers. The next range is a one followed by 227,832 zeros. This is the approximate size of the largest prime number known. As you can see the percent of primes at this range is a mere 19/100,000 of one percent or one prime out of every 525,000 numbers. The last range given is a one followed by a million zeros. What could possibly be this large? We have no reference. Here the number of primes is down to one in 2,300,000.

## SPECIAL FEATURES OF PRIMES

We know there is no workable formula for computing the $n$th prime. But if we know the value of a prime, then can we calculate the very next prime? In other words, if $P_n$ is the $n$th prime, then approximately where is $P_{n+1}$? If we know what $P_n$ is, can we say anything about the next prime, $P_{n+1}$? In 1845 Joseph Bertrand conjectured that if $n \geq 2$ then at least one prime exists between $n$ and $2n$. This is called Bertrand's observation, and was proved in 1850 by Pafnuti Tchebycheff. Therefore, if $P_n$ is a prime then there must be another prime less than $2P_n$. From this we know that the next prime, $P_{n+1}$, must be less than $2P_n$ or:

$$P_{n+1} < 2P_n$$

For example, if $n = 4$ then there will be a prime between 4 and 8. There is: 7. Now, if we know the value of a particular prime, we can at least always find a range where we can expect to find the next prime, even if that range may turn out to be rather large. To appreciate how large the range is, we only have to consider a prime such as 1,000,003. Bertrand's observation says that the next prime must be less than 2,000,006. This is not much help.

Joseph L. F. Bertrand (1822–1900) was a French mathematician who contributed to both number theory and the theory of probability in his book, *Calcul des probabilités*. Pafnuti Lvovich Tchebycheff (1821–1894) was a Russian who taught at the University of St. Petersburg. He was a rival of Lobachevsky, one of the cofounders of non-Euclidean geometry. While unaware of Carl Gauss' work, Tchebycheff helped prove the Prime Number Theorem.

We have two more relationships for $P_n$ and its successors. If $n \geq 2$ then $P_n + P_{n+1} > P_{n+2}$. This simply says that if you add a prime and its successor, you get a number greater than the next prime. For example, $5 + 7 = 12$ which is greater than the next prime, 11. Finally, we have: $P_m \cdot P_n > P_{m+n}$ or the product of two primes, $P_m$ and $P_n$, is greater than the $m+n$ prime. Thus, 7 is the fourth prime and 11 is the fifth prime. The above relationship says that $7 \cdot 11$ or 77 is greater than the $(4+5)$th or ninth prime. The ninth prime is 23 which is less than 77.

Now let's consider estimating the size of $P_n$. From the Prime Number Theorem it can be shown:

$$P_n \approx n \cdot ln(n) + n[ln(ln(n)) - 1]$$

Again we use the approximation sign ($\approx$) rather than the equal sign (=) to show that the above relationship only gives an estimate. The above equation[3] may look a mess but it is easy to compute since we really only have to compute $ln(n)$ and $ln(ln(n))$. Knowing $ln(n)$ we find the natural log of that number to get $ln(ln(n))$. Let's say that $n = 100$ and we want to compute the approximate size of the 100th prime: $ln(100) = 4.605$. This is substituted into $ln(ln(100))$ or $ln(4.605) = 1.527$. We can easily substitute these two values into the above equation.

$$P_n \approx (100)(4.605) + (100)[1.527 - 1]$$

Simplifying, we get $P_n \approx 513$. The 100th prime is exactly 541, so the approximation was off by 28 or a little over 5%. Table 8 gives examples of the above equation for different values of $n$ and the associated errors.

From Table 8 it is easy to see that the percent error decreases, on average, as $n$ increases. By the time we get to the 5000th prime the error is well below 1%. Even if we can't find the exact value for $P_n$ we can compute the size of $P_n$ to a high degree of accuracy as long as $n$ is relatively large. This can give us a close fix on the size of very large primes. For example, using the equation, we estimate the size of the one billionth prime to be 22,754,521,608 or just under 23 billion.

Considering our review of prime numbers, we realize at once that, while much is known about primes, much is still very hidden.

**Table 8.**  Estimated $P_n$ and Actual $P_n$

| $n$ | Estimate of $P_n$ | Actual $P_n$ | Difference | Percent Error |
|---|---|---|---|---|
| 500 | 3,521 | 3,571 | 50 | 1.40 |
| 1,000 | 7,840 | 7,919 | 79 | 1.00 |
| 1,500 | 12,454 | 12,553 | 99 | .79 |
| 2,000 | 17,258 | 17,389 | 131 | .75 |
| 2,500 | 22,203 | 22,307 | 104 | .47 |
| 3,000 | 27,260 | 27,449 | 189 | .69 |
| 3,500 | 32,409 | 32,609 | 200 | .61 |
| 4,000 | 37,638 | 37,813 | 175 | .46 |
| 4,500 | 42,937 | 43,051 | 114 | .26 |
| 5,000 | 48,296 | 48,611 | 315 | .65 |

In many cases this situation forces us to use equations involving only approximations. Knowing that prime numbers are the building blocks to the natural number sequence, and that the natural number sequence, in turn, forms the cornerstone to the great edifice of mathematics, we can appreciate why much work is continuing on unravelling the secrets behind these fascinating mathematical creatures.

## CHAPTER EIGHT

# PRIMES IN DEPTH

*And thus many are ignorant of mathematical truths,
not out of any imperfection of their faculties, or un-
certainty in the things themselves, but for want of application in acquiring, examining, and by due ways
comparing those ideas.*

JOHN LOCKE

*AN ESSAY CONCERNING HUMAN UNDERSTANDING*[1]

## A FISHING WE WILL GO

*E*xactly why millions of Americans go fishing each weekend during the summer is a mystery. Certainly, for the great majority, the cost of catching the few fish hooked far exceeds the value of the meat. In fact, when fishing some lakes and streams, we are required to throw back our catch, appreciating no gain to our palates or wallets. Why then do we do it? It's the fun of outsmarting the fish and either catching lots of fish, or catching the "big one" that got away last time. Once we get "hooked" on fishing, it's amazing to what extremes we'll go to find that special pond or lake loaded with native trout.

Searching for prime numbers is much like fishing. Mathematicians are not sure exactly why they do it, but they know it's great fun. Like the fisherman, we can go out looking for a whole bucket full of little, interesting primes, or we can go after that really big one. In fact, searching for primes is not the only compulsive behavior of which mathematicians are guilty. They are also known to become addicted to computing ever more digits in the decimal expansion of such numbers as π and $e$.

Prime numbers seem so easy to comprehend. Certainly, simple formulas must exist to help us compute them. Unfortunately, this is not the case. If we had perfect prime number formulas, one of them would enable us to easily compute the $n$th prime for any number $n$. Hence, we could just plug in 100 to get the 100th prime (which happens to be 541) and if we plugged in 1000 we would get the 1000th prime (7919).

While we do have a formula for calculating the $n$th prime, it's a very impractical formula to use. The mathematics needed to understand this formula is somewhat beyond this book. However, it is presented for those who are either disbelievers or possess a greater understanding of math. Before we can present the $n$th prime formula, we must investigate the special function, $F(j)$.[2]

$$F(j) = \left[ \cos^2\pi \, \frac{(j-1)! + 1}{j} \right]$$

To understand this interesting function, it is necessary to know the special convention regarding its surrounding brackets. If a number is surrounded by square brackets, [ ], then we will consider only the integral part of the number and disregard any fractional part. For example, [3.5] = 3 because we are going to ignore everything right of the decimal point. Thus, [1.999] = 1 and [0.638] = 0. Now for $F(j)$. The amazing characteristic of the function $F(j)$ is that it takes on the value of 1 when $j$ is a prime, but has the value of 0 when $j$ is a composite number. For example, we can test it using the prime 3 and the composite 4:

$$F(3) = \left[ \cos^2\pi \, \frac{2! + 1}{3} \right] = [\cos^2\pi] = [(-1)\,(-1)] = 1$$

$$F(4) = \left[ \cos^2\pi \, \frac{3! + 1}{4} \right] = \left[ \cos^2 \frac{7\pi}{4} \right] = \left[ \left( \frac{\sqrt{2}}{2} \right)^2 \right] = [0.5] = 0$$

This is truly an amazing function. How does it know when it's dealing with a prime or a composite? Notice what comes after the $\pi$ in $F(j)$, it's $j - 1$ factorial plus 1 divided by $j$. Remember Wilson's Theorem? We gave it as an example of an impractical way to test for primes, for the theorem says that $(j - 1)! + 1$ is divisible evenly by $j$ only when $j$ is a prime. That's how $F(j)$ knows when it's got a prime. When $j$ divides $(j - 1)! + 1$ evenly, then the number, $k$, is a whole number. But whenever this occurs, then $\cos^2 k\pi$ has the value of 1. Every time $j$ is *not* a prime, then dividing it into $(j - 1)! + 1$ leaves a fractional part. This causes the function $cos^2$ to have a value between 0 and 1. The brackets get rid of the fractional part and we're left with zero. What a wonderful use of Wilson's Theorem, and what a wonderful function in $F(j)$ to be able to spot a prime. In fact, we can now use $F(n)$ as a way to compute $\pi(n)$, our prime counting function. Since $F(n)$ is 1 when $n$ is prime and zero when $n$ is composite, we can add up all the $F(j)$ from 1 to $n$ as a count of the primes up to $n$. This yields one too many because 1 is not a prime, but $F(1) = 1$. Hence, we have:

$$\pi(n) = \sum_{j=1}^{n} F(j) - 1$$

Having defined $F(j)$ we are ready for the $n$th prime formula:

$$p_n = 1 + \sum_{m=1}^{2^n} \left[ \left( \frac{n}{\sum_{j=1}^{m} F(j)} \right)^{\frac{1}{n}} \right]$$

As you can appreciate, this formula is rather intimidating. Why isn't it useful? We have to have too much information to plug into it to get our answer. For example, to compute just the third prime (5) we must add eight different terms since m = 1 through $2^n = 2^3 = 8$. For each of these eight terms, we must take sums of $F(j)$. This step involves computing the factorial $(j - 1)!$ which can easily grow into a large number. If we are after the 100th prime, then $(100 - 1)!$ is a whopping:

933,262,154,439,441,526,816

992,388,562,667,004,907,159,682,643,816,214,685,929,638,952

175,999,932,299,156,089,414,639,761,565,182,862,536,979,208

272,237,582,511,852,109,168,640,000,000,000,000,000,000,000

With factorials growing so large, how would we ever manage to use the above equation to compute the millionth or billionth prime?

We can also give a formula to compute the next prime $P_{n+1}$ when we know the prime $P_n$, again based on the $F(j)$ function. In the following formula we will let $P_n = P$.

$$P_{n+1} = 1 + p + F(p + 1) + F(p + 1)F(p + 2) + \ldots + \prod_{j=1}^{p} F(p + j)$$

Again, the above formula requires too much information to be useful. Ideally, we would like a *simple* formula or equation to give us the $n$th prime. There is no such formula. It would also be nice to have a *simple* formula to yield the next prime. Again, no such formula exists. We must be willing to accept something less.

## FORMULAS THAT PRODUCE PRIMES

Many formulas do exist that produce nothing but prime numbers. These formulas do not produce each successive prime nor do they predict the next prime in sequence. But each time they are used, a prime pops out. However, as with the formulas in the previous section, they are only curiosities and of no real help in computing prime numbers. Our first example is actually rather simple. Again, we use the square brackets to mean we are only considering the whole number part of the number inside.

$$f(n) = [(1.3064)^{3^n}]$$

In this equation, $f(n)$ is the prime we get when we substitute the integer $n$ into the right side of the equation. After substituting in $n$ we compute the resulting number and then consider only the whole number part. For example, let $n = 1$. This gives us:

$$f(1) = [(1.3064)^{3^1}] = [(1.3064)^3] = [2.2296] = 2$$

Hence, the first number we get from this is 2 which is, indeed, a prime. Now let's substitute 2 for $n$.

$$f(2) = [(1.3064)^{3^2}] = [(1.3064)^9] = [11.0837] = 11$$

Again we get a prime, the number 11. Next we let $n = 3$.

$$f(3) = [(1.3064)^{3^3}] = [(1.3064)^{27}] = [1361.5332] = 1361$$

The number 1361 is also prime. However, at this point we can see what is happening. As $n$ increases, the values of $f(n)$ increase even faster. $f(4)$ is very large, a number with ten digits. $f(5)$ is a number with 29 digits. Hence, this formula increases in value so fast that it is of little use in generating a meaningful list of prime numbers.

Another formula that produces only primes is:

$$g(n) = [2^{2^{2^{.^{.^{.^{2^{1.92878}}}}}}}]$$

where the number of exponents to the first 2 is equal to $n$. This formula is a little more difficult to use. If we let $n = 1$ then:

$$g(1) = [2^{1.92878}] = [3.8073] = 3$$

Three is a prime number. Notice that when $n = 1$ we have only one exponent over the 2. If $n = 2$ then:

$$g(2) = [2^{2^{1.92878}}] = [2^{3.8073}] = [13.99976] = 13$$

Here again, we get a prime number. Now let $n = 3$:

$$g(3) = [2^{2^{2^{1.92878}}}] = [2^{13.99975}] = [16381.151] = 16{,}381$$

Again, we get a prime number. Yet we are beginning to generate primes so large they are hard to handle. $g(4)$ is a prime with 4932 digits. As can be seen, continuing with this formula would get us into numbers which are just too difficult to calculate. When using these two formulas, remember that the two constants used here, 1.3064 and 1.92878, are only approximations, since the exact values for these constants have not been precisely computed.[3]

We have seen formulas which always give us prime numbers, but they are of little use since they quickly explode into numbers too large to handle. Hence, they are of limited value.

Before we consider other formulas that produce primes, it is useful to introduce a helpful concept. All numbers can be factored into their prime numbers, of course. If two numbers, when factored, have no prime numbers in common then we say that they are prime *to each other* or that they are *relatively prime*. While 10 factors into 2·5, 21 factors into 3·7. Even though 10 and 21 are not prime numbers, they are prime to each other because they do not have a prime number in common. Consider the numbers 10 and 15 which factor into 2·5 and 3·5. They share the prime number 5 and, therefore, are not prime to each other.

Another way to say the same thing is to speak of the greatest common divisor (GCD) between numbers. The greatest common divisor of two numbers is just the product of all the prime numbers they share. Hence, the GCD of 10 and 15 is 5. No other number larger than 5 evenly divides both 10 and 15. If two numbers have no primes in common, then their GCD is 1.

> *Definition*: Two numbers are prime to each other (share no primes) when their GCD is 1.

Now we will consider a set of equations that do not always produce primes, but they still produce a high percentage of primes. We will consider first, second, and third degree polynomials which, remembering our high school algebra, are equations with the following forms:

First degree polynomial $\quad f(x) = AX + B$

Second degree polynomial $\quad f(x) = AX^2 + BX + C$

Third degree polynomial $\quad f(x) = AX^3 + BX^2 + CX + D$

In the above equations, $f(x)$ is the number (hopefully a prime) we are calculating. The $A$, $B$, $C$, and $D$ are coefficients, which are whole numbers, while $X$ takes on the value of successive whole numbers. We may be tempted to ask: Does there exist some polynomial that generates only prime numbers when substituting integers in for $X$? No. Unfortunately, every polynomial with integer coefficients will have infinitely many values of $X$ which produce composite numbers. Therefore, we can never hope to find a regular polynomial of any degree which will produce 100% primes. We have to settle for something substantially less.

We begin with the first degree polynomials which have the form $AX + B$. The number produced from the polynomial $AX + B$ can only be prime when $A$ and $B$ are prime to each other, in other words, when they are relatively prime. If $A$ and $B$ are not prime to each other, then $AX + B$ will be a composite number. If $A$ and $B$ are prime to each other then the resulting number might be prime. This is restated in a well-known theorem named after Peter Gustav Lejeune Dirichlet (1805–1859), who was a student of Carl Gauss.

> *Dirichlet's Theorem:* If $A$ and $B$ are prime to each other, then $AX + B$ is a prime for infinitely many values of $X$.

Dirichlet's fame, in part, comes from the fact that he helped make the work of the great mathematician, Carl Gauss, available to others. His 1837 proof of this theorem using higher mathematics of analysis demonstrated the remarkable fact that the mathematical fields of analysis and number theory are intimately connected.

Dirichlet's Theorem also gives us the following result: Since 1 is prime to every other number, then there must be an infinite number of primes of the form $An + 1$ where $A$ is any natural number.

We know from Dirichlet's theorem that if $A$ and $B$ are relatively prime (share no primes) then there exist an infinite number of primes of the form $AX + B$. Yet, this is not enough. What we want are specific polynomials which will produce a high number of primes.

If we now let $A = 2$ and $B = 1$ we get the polynomial $2X + 1$. This polynomial will generate the odd numbers or 1, 3, 5, 7, . . . (assuming we begin with $X = 0$). Hence, all but one prime (the number 2) will be in this sequence. How good is $2X + 1$ at generating primes? For the first 1000 terms in the sequence, we get 302 primes for 30.2%. As we continue substituting greater values into $2X + 1$, the percentage of primes we generate decreases because the overall density of primes decreases.

Another first degree polynomial producing a reasonable number of primes is $6X + 5$. Notice that 6 and 5 are relatively prime. This polynomial produces 39.6 percent primes for the first 1000 values of $X$. Another first degree polynomial producing numerous primes is $30X - 13$ which yields 41.1 percent primes for the first 1000 values of $X$. This is about as good as first degree polynomials get.

For improved prime generators we must look at the second degree polynomials having the form $AX^2 + BX + C$. One productive polynomial was discovered by Euler in 1772: $f(x) = X^2 + X + 41$. Here, the three coefficients ($A$, $B$, and $C$) are 1, 1, and 41 respectively. Let's see what happens when we let $x = 0$: $f(0) = 0^2 + 0 + 41 = 41$. Very good, we got the prime number 41. Let's try $x = 1$.

$$f(1) = 1^2 + 1 + 41 = 43$$

Nice! We got another prime. With excitement surging through our veins we look at Table 9 which shows $f(x)$ for $x = 0$ through 39. All 40 values of $f(x)$ are prime! Euler must have felt a great shock of excitement when he computed these first 40 solutions. But, things begin to fall apart as we continue computing $f(x)$, for the very next substitution with $x = 40$ gives us the composite number 1681 which is equal to 41·41. Therefore, our polynomial does not generate all primes, yet it does produce many primes.

Even though Euler's polynomial has begun to fail, it still produces primes at a high rate. For values of $x$ from 40 through 79, the polynomial produces 33 primes and only seven composite numbers. In the next 40 numbers (from 80 to 119) it generates 11

**Table 9.** First 40 Solutions for $X^2 + X + 41$, All of Which Are Prime

| $X$ | $f(X)$ | $X$ | $f(X)$ | $X$ | $f(X)$ | $X$ | $f(X)$ |
|---|---|---|---|---|---|---|---|
| 0 | 41 | 10 | 151 | 20 | 461 | 30 | 971 |
| 1 | 43 | 11 | 173 | 21 | 503 | 31 | 1033 |
| 2 | 47 | 12 | 197 | 22 | 547 | 32 | 1097 |
| 3 | 53 | 13 | 223 | 23 | 593 | 33 | 1163 |
| 4 | 61 | 14 | 251 | 24 | 641 | 34 | 1231 |
| 5 | 71 | 15 | 281 | 25 | 691 | 35 | 1301 |
| 6 | 83 | 16 | 313 | 26 | 743 | 36 | 1373 |
| 7 | 97 | 17 | 347 | 27 | 797 | 37 | 1447 |
| 8 | 113 | 18 | 383 | 28 | 853 | 38 | 1523 |
| 9 | 131 | 19 | 421 | 29 | 911 | 39 | 1601 |

composites, and from 120 to 159 we find 13 composites. If we test Euler's polynomial through the first 1000 values of $x$ we find that it produces primes 58.1% of the time. This is a reasonable improvement over the 41.1% primes we could get from the best first degree polynomial.

As E. Karst proved in 1973, the second degree polynomial producing the greatest number of primes for the first 1000 values of $X$ is $2X^2 - 199$, producing 598 primes.[4] When we use a first degree polynomial to generate numbers by substituting successive values for $X$ we get an arithmetic progression. An arithmetic progression, remember, is a sequence of terms where the difference between each term and its successor is a fixed or constant amount. What kind of number sequences do second degree polynomials produce? If we look at the successive values in Table 9 we find the first five differences to be: 2, 4, 6, 8, and 10. The difference between these five numbers is two. Hence, the *second* differences are constant. With the first degree polynomials, the first differences were constant and with the second degree polynomials the second differences are constant. This can be generalized to higher degree polynomials. If we have an $n$th degree polynomial, then the $n$th differences between terms will be constant.

What about third degree polynomials, which have the form $AX^3 + BX^2 + CX + D$? Unfortunately, little is known about the prime generating characteristics of polynomials of degree 3 and higher. By experimentation it is easy to find third degree polynomials which produce greater numbers of primes than a random search, yet none has so far produced more than Karst's or Euler's second degree polynomials for the first 1000 values of $X$. One interesting third degree polynomial is $X^3 + X^2 - 349$. This polynomial produces 411 primes for the first 1000 values of $X$, which is exactly what we got with the first degree polynomial $30X - 13$.

Table 10 shows the various polynomials we have considered along with the number of primes each generates for $X = 1$ to 1000, and the range of values for the polynomials for these 1000 values. Also shown is the percent of numbers which are prime numbers within that range.

**Table 10.** Comparison of Prime Generating Polynomials

| Polynomial | Number of Primes for $X$ = 1 to 1000 | Range of Polynomial Values | Percent Primes in Range |
|---|---|---|---|
| $2X + 1$ | 302 | 3 to 2,001 | 15.1 |
| $6X + 5$ | 396 | 11 to 6,005 | 13.0 |
| $30X - 13$ | 411 | 17 to 29,987 | 10.8 |
| $X^2 + X + 41$ | 581 | 41 to 1 Million | 7.85 |
| $2X^2 - 199$ | 598 | -197 to 2 Million | 7.45 |
| $X^3 + X^2 - 349$ | 411 | -347 to 1 Billion | 5.08 |

Why would we want polynomials that find concentrations of primes? Certainly, we already know the primes from 1 to a million or a billion. Yet, an intriguing question is whether equations exist of any type that are good at finding primes in great numbers at very high ranges of the natural numbers. Later we will learn the utility of huge primes, 75 or 100 digits long, in secret coding systems. A potential use of prime generating equations is to produce sets of large numbers that contain high percentages of primes. This could reduce the search time for individual primes.

## REALLY BIG PRIMES

It would seem that we humans just can't help ourselves: we are compelled to collect things. Collectable things might include the rarest major league baseball card, or the oldest beer can, or maybe just the record for the most goldfish swallowed. Mathematicians, it would seem, are not immune to this odd behavior. When it comes to prime numbers, they are constantly looking for the largest known prime. Of course, they are well aware that the primes are infinite, and therefore no largest one exists. But they can't help looking for one larger than the largest currently known. In fact, looking for larger primes is a good way to test new computers. The kinds of programs used to search for large primes are the very programs useful in testing new circuitry, for they require that the new computers complete huge numbers of sophisticated calculations in a short time. Several record-busting primes have been discovered using these programs.

Before his death in April of 1991, Samuel D. Yates of Delray Beach, Florida, served as a modern day Father Mersenne by collecting all known prime numbers with more than 1000 digits. He named such primes *Titanic primes*.[5] Many of these large primes have strange characteristics. For example, the 230th largest prime, with a total of 6400 digits, is composed of all 9s except one 8. That's 6399 9s and one 8! The 321st prime, with 5114 digits, is composed of only 1s and 0s. The 41st prime (11,311 digits) is a palindrome, reading the same forward or backward. The 297th prime (5323 digits) has a single 4 followed by 5322 nines.

Certainly one of the strangest primes is the 713th largest prime or $(10^{1951}) \cdot (10^{1975} + 1991991991991991991991991) + 1$. This odd number, with 3927 digits, was discovered by Harvey Dubner in what year? You guessed it—1991!

These large primes attracted enough attention that Chris K. Caldwell at the University of Tennessee maintains a list of Samuel Yates' Titanic primes on the World Wide Web on the Internet. This file has grown to over 9000 primes, with a new category for primes over 5000 digits called *Gigantic primes*.

The second to the largest prime, discovered in 1992 by David Slowinski and Paul Gage, is $2^{756839} - 1$. With 227,832 digits it looked at the time to be an enduring contender for the crown. It was discovered after David Slowinski of Cray Research Inc. of Chippewa Falls, Wisconsin, developed a computer program specially designed to search for a special kind of prime called a Mersenne prime. Scientists at AEA Technology Harwell Laboratory in Harwell, England, decided to use Slowinski's program to test a new Cray computer. Slowinski provided the Harwell team with 100 large numbers he thought might be Mersenne primes. After only 19 hours of computation, the computer kicked out the second largest prime ever discovered. Here is one good example of a prime being discovered while testing a new computer.

But the above prime is only the second largest. We are now ready for what you have been holding your breath for—the biggest prime known (a big drum roll, please). This is it—Number One, the record prime until someone discovers a bigger one.

## The Gargantuan Biggest Prime = $2^{859433} - 1$

This prime has a total of 258,716 digits, and was also discovered by the team of Slowinski and Gage in 1994. To fully appreciate just how big these last two primes are, we can compare them to the third largest prime which is $391581 \cdot 2^{216193} - 1$. This number, discovered in 1989 by a team of prime hunters, has only 65,087 digits. That's only approximately one fourth the number of digits of our two heavyweights.

You might be asking by now: What's the big deal? Why would we want to find big primes anyway? We have already mentioned the testing of computers, but the real reason goes far beyond this practical application. First, the ability to find and verify such large primes is a yardstick on our ability to develop both hardware and computational routines. The attempt to find even larger primes may uncover new and productive number crunching techniques. However, even this is not the primary reason. For this we must simply point to the indomitable nature of men and women to excel in their pursuit of excellence. If there is a higher mountain to climb, we must climb it. If there is a larger prime to find, by gum, we're going to find it! By the time you read this sentence, the record for the largest prime will, in all likelihood, have already been broken, for not only are individual mathematicians calculating away for the record, but whole teams of mathematicians with sophisticated computer programs are in the competition.

### SPECIAL KINDS OF PRIMES

Now it's time to look at some special kinds of prime numbers. Of course, the most unique prime number has already been discussed: It's the number 2, the only even prime. Oddness and evenness is understood by all of us at an early age, and divides all numbers into two groups. All even numbers with the exception of 2 are composite numbers. Therefore, to study prime numbers is to study odd numbers and some of our special primes are, indeed, very odd.

Twin primes are pairs of primes of the form $P$ and $P + 2$. All twin primes differ by only two numbers. Many examples of twin primes

**Table 11.** Number and Percent of Twin Primes

| $n$ | Number of Primes | Number of Twin Primes | Percent Twin Primes |
|------|------------------|----------------------|---------------------|
| $10^3$ | 168 | 35 | 20.83 |
| $10^4$ | 1,229 | 205 | 16.68 |
| $10^5$ | 9,592 | 1,224 | 12.76 |
| $10^6$ | 78,498 | 8,169 | 10.41 |
| $10^7$ | 664,579 | 58,980 | 8.87 |
| $10^8$ | 5,761,455 | 440,312 | 7.65 |
| $10^9$ | 50,847,534 | 3,424,506 | 6.73 |
| $10^{10}$ | 455,052,512 | 27,412,679 | 6.02 |
| $10^{11}$ | 4,118,054,813 | 224,376,048 | 5.45 |

exist in the smaller numbers: 3 and 5, 5 and 7, 11 and 13. An example of a substantially larger pair is 55,049 and 55,051. Is there an infinite number of twin primes? No one knows! However, as the natural numbers get larger, the percent of primes which are twin primes decreases. Look at Table 11. While approximately 20% of the smaller primes are also a member of twin primes, this percentage decreases among the larger primes. It is not presently known whether twin primes thin out until they disappear entirely. The largest known twin primes are $697053813 \cdot 2^{16352} + 1$ and $697053813 \cdot 2^{16352} - 1$, which are numbers with 4932 digits each, found by Indlekofer and Ja'rai in 1994.

Remember when we formed the infinite series of all the reciprocals of prime numbers? That series, we know, is unbounded. But what about the series formed from the reciprocals of all twin primes? Certainly, the twin prime series is less dense than the series using all the primes, since the fraction of primes that are twins diminishes quickly once we get into larger and larger numbers. In fact, we really don't know if such a series would even be infinite, since we don't know if there exist infinitely many twins.

$$\sum_{p = twin\ primes}^{?} \frac{1}{p} = \frac{1}{3} + \frac{1}{5} + \frac{1}{7} + \frac{1}{11} + \frac{1}{13} + \cdots$$

Even though mathematicians don't know whether this series has an infinite number of terms, they do know that it converges. In

1919 V. Brun proved it converged, and then went on to prove that for any large number $N$, somewhere in the number sequence exists a succession of $N$ primes that are not twins. This is similar to the idea that for any number, $N$, we can find a succession of $N$ composite numbers. The value that the reciprocal twin sum converges to is called Brun's Constant and has been calculated to approximately 1.90216. We don't really know what kind of number this is: Is it rational? Algebraic? Transcendental? Knowing that the series of reciprocal twin primes converges while the reciprocal of all primes diverges tells us that twin primes are relatively scarce compared to all primes even if they turn out to be infinite in number.

## MERSENNE AND FERMAT PRIMES

Father Marin Mersenne (1588–1648) was a French Minimite friar who was also an amateur mathematician. He acted as a kind of post office for other mathematicians of his day including Pierre de Fermat. Besides his involvement in mathematics, he showed the relationship between the period of vibration of a violin string and its density, tension, and length. He is known to have defended both René Descartes and Galileo Galilei against theological criticism.

Mersenne numbers have the form $M_p = 2^p - 1$ where $p$ is a prime number. Hence, the following numbers are Mersenne numbers: $2^2 - 1 = 3$, $2^3 - 1 = 7$, and $2^5 - 1 = 31$. Notice that these three numbers, 3, 7, and 31, are primes. Not all Mersenne numbers are primes. Mersenne $M_{11}$ is composite for $2^{11} - 1 = 2047 = 23 \cdot 89$. In fact, to date there are only 32 known Mersenne primes. How many Mersenne numbers are prime and how many are composite? No one knows. There may be an infinite number of both kinds.

Historically, Mersenne primes have been important, often holding the record as the largest primes known. Currently the largest and third largest primes are Mersenne numbers. The fourth largest prime is the Mersenne prime $2^{216,091} - 1$, which has 65,050 digits and was discovered in 1985 by David Slowinski. The exponents of 2 which might yield a Mersenne prime have been searched from 216,091 to 355,031 and again from 430,000 to 524,287 without revealing any new Mersenne primes. A select group of computer

FIGURE 37.   Pierre de Fermat, 1601–1665.

scientists, calling themselves the "Gang of Eight," continues the quest!

Pierre de Fermat (1601–1665) was possibly the world's greatest amateur mathematician and responsible for founding modern number theory (Figure 37). Fermat conjectured that numbers of the form

$$F_n - 2^{2^n} + 1$$

are always prime. Such numbers are now called Fermat numbers. We now know that Fermat's conjecture is false. Some Fermat numbers are prime and some are not. Following is a list of the first seven Fermat numbers:

| | |
|---|---|
| $F_0 = 3$ | prime |
| $F_1 = 5$ | prime |
| $F_2 = 17$ | prime |
| $F_3 = 257$ | prime |
| $F_4 = 65,537$ | prime |
| $F_5 = 641 \times 6,700,417$ | composite |
| $F_6 = 274,177 \times 67,280,421,310,721$ | composite |

As you can see, the Fermat numbers grow in size very quickly. The Fermat numbers $F_7$, $F_8$, $F_9$, and $F_{11}$ have all been factored and are composite. The largest known Fermat prime is $F_4$. Pierre de Fermat didn't have a table of primes large enough to see that $F_5$ was composite. A century later Leonhard Euler discovered the factors to $F_5$ and demonstrated that it was composite. The largest known Fermat number is $F_{23471}$ which has $10^{7000}$ digits.

Are there an infinite number of Fermat primes? Composites? As with the Mersenne numbers, no one knows.

## GOOFY PRIMES

A *primorial* number has the following form: $p\# + 1$ where $p\#$ is the product of all the primes less than or equal to $p$. Hence, $3\# + 1 = 2 \cdot 3 + 1 = 7$. Therefore, the primorial number $3\# + 1$ is a primorial prime. Not all primorials are primes since $13\# + 1 = 2 \cdot 3 \cdot 7 \cdot 11 \cdot 13 + 1 = 30,031 = 59 \cdot 509$. The largest known primorial prime is $24029\# + 1$ which has 10,387 digits, and was discovered by Chris Caldwell in 1993.

A *repunit* is a number with all the digits equal to 1. Hence, 1, 11, 111, and 1111 are all repunits. We generally identify repunits as $Rn$ where $n$ gives the number of 1s. The only known prime repunits are $R2$, $R19$, $R23$, $R317$, and $R1031$[6]. If $Rn$ is a prime then $n$ must also be a prime. But, just because $n$ is prime does not guarantee that the resulting $Rn$ is prime. Are there infinitely many prime repunits? No one knows.

Some primes, called palindromes, read the same from the left or the right. Hence, 11 and 10,301 are both palindromes. The largest known palindrome, with 11,311 digits, is:

$$10^{11,310} + 4,661,664 \times 10^{5652} + 1$$

Discovered in 1991 by Harvey Dubner, it is the 41st largest prime.

Some Fibonacci numbers are primes. The largest known to date is $FN(2971)$ (that is, the 2971st term of the sequence), discovered by Hugh C. Williams.[7] Are there infinitely many Fibonacci primes? No one knows.

Another strange prime is the Sophie Germain prime. Mademoiselle Sophie Germain (1776–1831) was a Frenchwoman who contributed to the theory of acoustics, the mathematical theory of elasticity, and higher arithmetic. Afraid that the great mathematician Carl Gauss would be prejudiced against a woman mathematician, she corresponded with him under the name of Mr. Leblanc. Yet, when her real identity became known, Gauss encouraged and delighted in their extended correspondence. Unfortunately, they never met in person, and she died before receiving an honorary doctor's degree, recommended by Gauss, from the University of Göttingen.

A prime number $p$ is a Sophie Germain prime if you can double it, add 1 and get another prime, i.e., if $p$ is prime and $2p + 1$ is also prime. The smallest Sophie Germain prime is 2 because $2 \cdot 2 + 1 = 5$ which is prime. The next is 3 since $2 \cdot 3 + 1 = 7$. The largest known Sophie Germain prime is: $9402702309 \cdot 10^{3000} + 1$ (hence, twice this number plus 1 is also a prime) which has 3010 digits and was discovered by Harvey Dubner in 1993. It is unknown whether there are an infinite number of Sophie Germain primes.

A rather obscure kind of prime is the Wilson prime. The prime $p$ is a Wilson prime when $(p - 1)! + 1$ is evenly divisible by $p^2$.

The smallest Wilson prime is 5 for $(5 - 1)! = 2 \cdot 3 \cdot 4 = 24$. If we add 1 to 24 we get 25 which is divisible by $5^2$. There are only three known Wilson primes: 5, 13, and 563. No one knows if there exists an infinite number.

As the cliche goes, we have only scratched the surface when it comes to strange primes. Mathematicians, in their attempt to know and understand prime numbers, have defined many more interesting types. For example, there are not only Fermat primes, but generalized Fermat primes. There are tetradic, pandigit, and prime-factorial plus one primes. And there are Cullen, multifactorial, beastly palindrome, and antipalindrome primes. Add to these the strobogrammatic, subscript, internal repdigit, and elliptic primes. In fact, a whole new branch of mathematics seems to be evolving that deals specifically with the attributes of the various kinds of prime numbers. Yet understanding primes is only part of our quest to fully understand the number sequence and all of its delightful peculiarities.

## LINE THEM UP IN ROWS AND COLUMNS

We can do some fun things with numbers if we arrange them in rows and columns. Doing so, we find that some kinds of numbers fall into certain columns. This happens with prime numbers. Look at Table 12. Here we have arranged the numbers in six

**Table 12.** A Six-Column Array of Numbers
(primes are marked)

| Column 1 | Column 2 | Column 3 | Column 4 | Column 5 | Column 6 |
|----------|----------|----------|----------|----------|----------|
| 1 | **2** | **3** | 4 | **5** | 6 |
| **7** | 8 | 9 | 10 | **11** | 12 |
| **13** | 14 | 15 | 16 | **17** | 18 |
| **19** | 20 | 21 | 22 | **23** | 24 |
| 25 | 26 | 27 | 28 | **29** | 30 |
| **31** | 32 | 33 | 34 | 35 | 36 |
| **37** | 38 | 39 | 40 | **41** | 42 |
| **43** | 44 | 45 | 46 | **47** | 48 |
| 49 | 50 | 51 | 52 | **53** | 54 |
| 55 | 56 | 57 | 58 | **59** | 60 |
| **61** | 62 | 63 | 64 | 65 | 66 |
| **67** | 68 | 69 | 70 | **71** | 72 |
| **73** | 74 | 75 | 76 | 77 | 78 |
| **79** | 80 | 81 | 82 | **83** | 84 |
| 85 | 86 | 87 | 88 | **89** | 90 |
| 91 | 92 | 93 | 94 | 95 | 96 |
| **97** | 98 | 99 | 100 | **101** | 102 |
| **103** | 104 | 105 | 106 | **107** | 108 |
| **109** | 110 | 111 | 112 | **113** | 114 |
| 115 | 116 | 117 | 118 | 119 | 120 |

columns beginning with the number 1 in the first row, first column. We have also highlighted those numbers that are primes. Notice that in Table 12, all the primes, except for 2 and 3, are either in column 1 or 5. What's going on? It turns out that all the infinity of prime numbers, except 2 and 3, fall either in the first or fifth column. In other words, after the first row, no primes exist in the other four columns. We can express this algebraically by saying that all primes larger than 3 have the form $6n + 1$ or $6n + 5$, where $n$ is some integer. Unfortunately, there are also many composite numbers with this form.

When mathematicians talk about numbers being in various rows and columns of a number array, they use a special language. If two numbers, $A$ and $B$, share the same column when all the natural numbers are placed in an array with m columns, then they say that "$A$ is congruent to $B$ modulo $m$." This is written symbolically as: $A \equiv B \bmod m$ (but it really means $A$ and $B$ are in the same column.) Carl Gauss developed this way of relating numbers. For example, we can look again at Table 12 and see that 34 and 16 are in the fourth column. Therefore, we can write $34 \equiv 16 \bmod 6$, and we say that 34 is congruent to 16 modulo 6. Notice that if we divide both 34 and 16 by 6 we get the same remainder, or 4. This is always the case, i.e., when $A$ is congruent to $B \bmod m$, then we get the same remainder when we divide both $A$ and $B$ by $m$.

The congruence relationships between numbers has grown into an entire branch of mathematics known as congruence theory. Congruence theory can be very helpful in solving certain problems involving large numbers. This is because some solutions depend more on which column a number falls into, rather than its absolute magnitude. We will see an example of how congruence arithmetic is used to solve a problem when we investigate numbers used in secret codes.

Thinking of numbers in columns and using the congruence notation can be of great help with certain problems, and can help us in understanding prime numbers. For example, we can now say that all primes larger than 3 are either congruent $1 \bmod 6$ or congruent $5 \bmod 6$; that is, will either leave a remainder of 1 or 5 when divided by 6.

We should look at Table 12 one more time for there are still secrets lurking within. Remember when we talked of two numbers being relatively prime to each other? $A$ and $B$ are relatively prime when they have no common factors. Now, notice that both 1 and 5 are relatively prime to 6 while the other column numbers are not: 2, 3, and 4 are not prime to 6, but share at least one factor. This also can be generalized into a rule. For any array with $m$ columns, the primes beyond the first row will only occur in those numbered columns that are prime relative to $m$. For example, in an array of seven columns we can say that primes will occur in the first six columns since the numbers 1, 2, 3, 4, 5, and 6 are all prime relative to seven. However, in an 8-column array, the primes beyond the first row will only occur in columns 1, 3, 5, and 7 since these numbers are all relatively prime to 8.

We now have a nice relationship between the location of prime numbers in arrays, and those numbers which are prime to $n$ and less than $n$. Such numbers, in fact, identify the columns where primes occur. We actually have a name for a function that counts the number of numbers less then $n$ and prime to $n$. It is called *Euler's* $\phi$-function and is designated as $\phi(n)$. (This use of $\phi$ should not be confused with the $\phi$ used to designate the Golden Mean. The two are entirely different.) We have already seen that $\phi(6) = 2$ since two numbers less than 6 are prime to 6 (1 and 5), and $\phi(7) = 6$ (1, 2, 3, 4, 5, 6) while $\phi(8) = 4$ (1, 3, 5, 7). In order to use $\phi(n)$ in general formulas where the index may be 1, we define $\phi(1) = 1$.

We have seen how various formulas can be used to find the number of primes less than a certain number, and we have studied other formulas that generate sequences of prime numbers. However, we are left with a sense of dissatisfaction with such formulas because they are really curiosities, and not useful in computations. This leads to the natural conclusion that we still have much more work to do in solving the riddles surrounding prime numbers. The search for large primes and special kinds of primes reinforces our desire to gain a better grasp of how primes are distributed in the natural number sequence.

## CHAPTER NINE

# PRIMES AND SECRET CODES

*. . . but as the roads between Media and Persia were*
*guarded, he had to contrive a means of sending word*
*secretly, which he did in the following way. He took a*
*hare and cutting open its belly without hurting the*
*fur, he slipped in a letter containing what he wanted*
*to say, and then carefully sewing up the paunch, he*
*gave the hare to one of his most faithful slaves, dis-*
*guising him as a hunter with nets, and sent him off*
*to Persia . . .*[1]

HERODOTUS (440 B.C.)

## A BRIEF HISTORY OF SECRET CODES

*H*umankind's desire to conceal sensitive messages is only surpassed by its ingenuity in devising the means to do so. Properly speaking, the secret to be sent is called a *message* (also called the *plaintext*). The sender alters this message by *enciphering* it or creating a *ciphertext* which cannot be read by any unauthorized person (an intruder or spy) who may intercept it. This cipher is transmitted to the receiver who *deciphers* the ciphertext, changing it back into the original message. This whole science is called *cryptography*. An unauthorized person who wishes to understand the message must *break* the cipher. Breaking ciphers is the science of *cryptanalysis*.

The importance of maintaining state secrets through cryptography cannot be overstated. In past wars, great battles have been decided because an enemy was or was not able to "crack" a secret code. The ability of American cryptanalysts to break the Japanese naval code *JN25* during World War II was pivotal to the American victory over the superior Japanese fleet at the battle of Midway. Governments have spent huge amounts to break other countries' codes to steal their secrets. The National Security Agency (NSA) of the United States owns rooms full of modern computers used to study secret codes. Yet, because the ultimate goal of cryptography is to keep secrets, there is not a great body of literature on cryptographic techniques. A modern twist to the spy game is industrial espionage, where one international company tries to steal trade secrets from it competitors.

As early as 450 B.C. the Spartans of Greece used a cipher for sending messages. A belt, called a scytale, was wrapped around a wooden cylinder. The message was then written on the belt. Unwound and worn by the courier, the belt was just a jumble of marks, but when rewound on a similar cylinder by the receiver the belt revealed the desired message. Julius Caesar supposedly used letter substitution for his secret writing. Yet, little mention of codes or ciphers can be found for a thousand years after Greece's late classical period of 100 A.D. During the 13th century, Roger Bacon described several coding systems and completed a book written in cipher which has never been broken. The first modern work on cryptography was written by a German abbot, Johannes Trithemius, in 1510.

The ability of radios to quickly transmit orders to soldiers in the field or diplomatic instructions to embassies generated a great need for 20th century governments to develop good ciphers. This resulted in a technological explosion of means for sending messages as well as breaking them.

Two kinds of classical ciphers exist: transposition and substitution. In the transposition cipher, the letters comprising the message are scrambled according to a specific procedure. The receiver then unscrambles the letters by reversing the procedure. One such transposition cipher is the *route cipher* where the sender makes a rectangle out of the letters of the message. He then chooses a route through the rectangle which scrambles the letters. The receiver can reconstruct the proper rectangle since he or she knows the route.

One of the most common transposition cipher techniques used during World War II was a ciphering machine which had 26 disks, each rotating independently of the others around a common axis. On the surface of each disk was a complete, but scrambled, alphabet. The disks were rotated until part of the message could be read across the disks. The cipher for that part of the message was then read across the disks at a different location. The German version of this machine was called the *Enigma*, one of which was obtained by the British. The *Enigma*, in conjunction with cryptanalysis techniques, allowed the British to decipher many German messages, which significantly altered World War II's outcome.

With the substitution cipher, letters or groups of letters of the message are replaced with other letters or symbols. However, a simple substitution cipher is relatively easy to break by an intruder. Every language has certain characteristics which, when understood by the intruder, makes it possible to guess at specific letters in the cipher. For example, the letter E is the most common letter in English, French, Spanish, and German. Knowing this, the intruder substitutes E into the most frequent letter or symbol of the cipher. Next he or she tries other letters. If the enciphered message is sufficiently long, this type of analysis will frequently break the code.

Many 20th century ciphers have been based on a *two-step* cipher. In this technique, the message is altered by both a transposition and a substitution method. First, the message is changed by substituting a two-digit number for each letter. This allows for sending the message in digital code over various kinds of telecommunications lines including telephone, radio, and microwave transmitters. After the letters are changed into numbers, the numbers are scrambled.

The most secure cipher ever devised is the two-step cipher known as the *one-time sheet* or *one-time tape*. This system uses a book consisting of pages filled with random digits. No two pages repeat the same sequence of numbers. The sender first substitutes two-digit numbers for each letter in the message. Then a sequence of random numbers is taken from one page (sheet) of the book and subtracted from the message digits. This produces a random sequence of numbers, and an intruder cannot use any common technique to break the code. Once used, a sheet is torn out of the book and discarded to insure that particular sequence of numbers will never be used again on a message. The cipher's receiver uses the same page in his or her book to add back the random digits and regenerate the original message.

Since there are no repeating patterns in a one-time sheet cipher, an intruder must either steal a copy of the random number book, or steal a copy of the message before enciphering or after deciphering. A tough job. The real drawback with this cipher system is that the sender must constantly produce random number books and get them securely delivered to the receivers. For high volume communications, this becomes a daunting problem.

One current popular cipher is the Data Encryption Standard (DES) which is used to encipher computer messages.[2] The cipher began as the Lucifer cipher developed by IBM but was released by the computer giant in the mid-seventies to become the standard for banks and government agencies, who needed to keep their transactions confidential. In 1977 the U.S. government adopted DES as an official standard for data encryption. The cipher uses a key of 56 binary bits. Since there are 72 quadrillion ways to assign 56 bits, individual keys are relatively safe from code breakers. For example, a home computer would require tens of thousands of years to test all possible combinations in just one 56-bit key.

## THE PUBLIC-KEY CODE

All of the classical cipher systems, including the DES system, suffer from two serious defects. First, the receivers must posssess a secret key for deciphering received messages. How do you deliver these keys to dozens or hundreds of potential users without sending out couriers to hand deliver them? Another problem involves message signatures. As receivers, how do we know a message is authentic since, with modern telecommunications, we don't actually "see" a physical signature. When we download data or programs into our computers, how do we know it has not been tampered with and is virus-free? Having a signature is also important when messages contain legal records or financial transactions.

A new kind of code, called a public-key code, solves both these problems. A receiver makes public an enciphering key for all to see. Using this key, anyone can be a sender, enciphering a message and sending the ciphertext to the receiver. The receiver has a secret *deciphering key* which he or she uses to decipher the ciphertext and produce the message. "How can this be?" you say. "If I can encode a message, then I can just reverse the procedure and decode the message." But this is not true with a public-key code. With it, I can encipher a message if I have the enciphering key but can't decipher it without the deciphering key. The enciphering key works in only one direction: to encipher messages. Since the enciphering key is openly published, there is no need for the receiver to send couriers with secret keys to each potential sender. It is also impossible for senders to accidently or deliberately reveal any deciphering key to would-be intruders.

The mechanism for achieving the encoding and decoding depends on a very large composite number that factors into only two large prime numbers. The sender uses the composite number to encode the message. The receiver uses the two prime number factors to decode it. The process's security depends on the sender's inability (or the public's) to factor the composite number.

Through an ingenious device, the sender can also "sign" the message in such a way that uniquely identifies it. Thus, the public-key code solves the second problem of signatures. Both the sender and receiver have their own public and secret keys. If the sender wishes to attach his signature to a message, he simply encodes his name with *his* secret key. The receiver can then apply the sender's public key to read the sender's name. The message must have come from the sender, since only the sender possesses his secret key which will encipher his name so that his public key will decipher it.

Solving these two problems makes the public-key cipher perfect for commercial uses. In this modern day of computers and high speed communications, businesses need to transmit large volumes of secret, secure data to subsidiaries, suppliers, head offices, and customers. Using the public-key cipher requires only that the receiver make public his or her enciphering key. This cipher system is new, yet it is currently being adopted as the standard cipher system of the future. In 1976 Whitfield Diffie and Martin Hellman outlined the mathematics for a public-key system.[3] Based on their ideas, a commercial encryption system was invented by Ronald Rivest, Adi Shamir, and Leonard Adleman in 1977.[4] In 1982 the inventors founded RSA Data Security, Inc. of Redwood City, California, to market the system.

## HOW IS IT DONE?

The public-key cipher system is based on the mathematics of prime numbers, and how numbers distribute themselves in rectangular arrays. The system uses both transposition and substitution. First each alphabetic letter from the original message is assigned to a two-digit number—this is substitution. Next, mathematics is used to scramble the two-digit numbers. This is transposition. To understand how the numbers are scrambled look at Table 13 which

**Table 13.**  Five-Column Array

| Column 1 | Column 2 | Column 3 | Column 4 | Column 5 |
|----------|----------|----------|----------|----------|
| 1        | *2*      | 3        | *4*      | 5        |
| 6        | 7        | *8*      | 9        | 10       |
| 11       | 12       | 13       | 14       | 15       |
| *16*     | 17       | 18       | 19       | 20       |
| 21       | 22       | 23       | 24       | 25       |
| 26       | 27       | 28       | 29       | 30       |
| 31       | *32*     | 33       | 34       | 35       |
| 36       | 37       | 38       | 39       | 40       |

is just a five-column array of numbers. In it we have placed asterisks next to each power of the number 2. These include the numbers 2, 4, 8, 16, and 32. Notice that while 2 is in column 2, 4 moves to column 4. Eight is in column 3, 16 is in column 1, and finally 32 is back in column 2 again. This movement of powers of numbers to different columns repeats itself. In our example, 2 raised to the 5th power ($2^5 = 32$) ended back in the same column as the original number 2. Without completing the table we can predict that the next power of 2, that is 64, will be in column 4, while the following power, 128 will be in column 3. Notice that 64 divided by 5 leaves a remainder of 4 (column 4) and 128 divided by 5 leaves a remainder of 3 (column 3). Remember that identifying the columns that numbers fall into in an array is called congruent mathematics or the study of congruences. When $A$ and $B$ share the same column in an $m$ array, we show this symbolically as $A \equiv B \bmod m$, where *mod* stands for modulo. Another way of thinking of the relationship between $A$ and $B$ is that they both leave the same remainder when divided by $m$.

Now let's demonstrate how the characteristic of powers repeating in columns can be used to scramble and unscramble numbers in a cipher. First we write the numbers into a five-column array such as Table 13. Next we assign the alphabetic letters to various columns. Suppose that we assign the letter $A$ to column 2 and the letter $B$ to column 3. To encipher our $A$ (which is now the two-digit number 02) we could choose to raise 02 to its third power or $2^3$. This gives us the number 8 which falls in column 3. Because $2^3$ or 8 is in column 3 we change the 02 (representing $A$) into an 03 (representing $B$). Hence, 02 is enciphered to be a 03 and $A$ becomes a $B$. If someone tried to read the message (and knew how we assigned numbers to letters) they would see the 03 and think it was a $B$.

How does the receiver of this ciphered 03 get it back to a 02 and its corresponding $A$? The receiver has the secret deciphering key which is to multiply the 03 by the proper power of 2 which in this case is $2^2$ or 4. Why would he or she do this? Look again at Table 13. When we raised 2 to the third power we got 8. If we multiply 8 by 4 we get 32 which is back in column 2. Because of the laws of congruences, we can also multiply 3 by 4 to get 12 which is also back in column 2. In fact, any number in column 3 multiplied by 4 gets us back to our original column, 2. Therefore, using the deciphering key of 4 the receiver multiplies 03 to get 12. Twelve is back in column 2, so the proper letter to decipher is $A$. To recap: The sender raises 02 to its third power giving us 8, which is in column 03 (= $B$). To decipher the 03 ($B$) back to 02 ($A$) the receiver multiplies the 03 by 4 to get 12, which is in column 2 (= $A$).

In order to encipher a number, the sender must know two things: the correct power to raise the original number, and how big an array is being used so that he or she can reduce this number to an appropriate number in the first row. The receiver must also know the secret number with which to multiply the cipher and return it to the correct column. Of course, this example is exceedingly simplified. When we construct our actual cipher we must make sure that giving the enciphering key to the public (the potential senders) does not accidentally give away the secret deciphering key.

To make the whole procedure work we must be able to find the correct power to use when raising a number from the first row of an array so that the resulting number will return to that correct column. For that we introduce an elegant theorem by Euler which deals with powers of numbers and arrays.

> *Euler's Theorem:* If $n$ and $m$ are relatively prime, then when we raise $n$ to the $\phi(m)$ power, the resulting number will be in column #1 of an $m$-array. [If $(n,m) = 1$ then $n^{\phi(m)} \equiv 1 \pmod{m}$.]

For convenience, we state the theorem in everyday English. The notation following in brackets shows how the theorem might look in an elementary number theory book. In this theorem, $\phi(m)$ is Euler's function, which is the number of numbers less than $m$ and prime to $m$. Hence, $\phi(m)$ counts the number of numbers less than $m$ that share no primes with $m$. Euler's Theorem places the original number, $n$, back into column 1 by raising it to the appropriate power. But for our secret code we want the number $n$ to return to its original column and not column 1. To achieve this, we are going to alter Euler's theorem to use in our public-key cipher. By applying the laws of congruences we can multiply both sides of the congruence equation in the theorem by $n$ and change the Euler theorem to the following:

> *Theorem*: If $n$ and $m$ are relatively prime, and $k$ is any integer, then when we raise $n$ to the $k \cdot \phi(m)+1$ power, the resulting number will be in column $n$ of an $m$-array. [If GCD$(n,m) = 1$ then $n^{k\phi(m)+1} \equiv n \pmod{m}$.]

This theorem says that if we take any integer $k$, multiply by $\phi(m)$, and add 1 we get the desired power of $n$ to get back to column $n$. This gives us the ability to predict just when the power of a number $n$ will return it to the column that $n$ started in.

Let's take another example. This time we will start with a six-column array of numbers. We'll let $n$ be in the fifth column since 6 and 5 are relatively prime. Now $\phi(6) = 2$ (there are only two numbers less than 6 which are relatively prime to 6, namely 1 and 5). So if we take 5 to the $2k + 1$ power we should get another number in the fifth column. Let's see what happens when $k = 1$. Substituting, we get $k \cdot \phi(m) + 1 = 1 \cdot 2 + 1 = 3$. Five to the third power is 125. One hundred twenty-five divided by 6 leaves a remainder of 5, and we get a number in the fifth column. It worked. Let's try $k = 2$. This gives us $k \cdot \phi(m) + 1 = 2 \cdot 2 + 1 = 5$. Five raised to the fifth power is 3125. Dividing by 6 again, we get a remainder of 5. Hence, 3125 is also in the fifth column. Therefore, whenever we raise 5 to the $2k + 1$ power, we get a number which is back in column 5.

## AN EXAMPLE

We will now review a complete example of how the system works. Let's begin with a 55-column array. The number 55 factors into $5 \cdot 11$. We can compute $\phi(55)$ according to the following simple rule: if $n = p \cdot q$, where $p$ and $q$ are two primes, then $\phi(n) = (p - 1) \cdot (q - 1)$. Therefore,

$$\phi(55) = (5 - 1)(11 - 1) = 4 \cdot 10 = 40$$

Now we know that any column number prime to 55 which is raised to the $40k + 1$ power will give a number back in that same column. We will let $k = 4$ so that $k\phi(55) + 1 = 4 \cdot 40 + 1 = 161$. Now, all the numbers in the first row of a 55-column array which are relatively prime to 55 when raised to the 161th power will result in a number in the original column.

The number 161 factors into $7 \times 23$. One of these two numbers will be our public enciphering key and the second will be our secret deciphering key. Let 7 be the public key and designate it as $E$. The deciphering key will be 23 and designated as $D$. Hence: $E \cdot D = k \cdot \phi(m) + 1$ or $7 \cdot 23 = 4 \cdot 40 + 1 = 161$.

We are going to select four columns in our 55-column array which are all prime to 55. Let's choose columns 2, 3, 4, and 6. We will assign the following letters to these numbers: $H = 02, E = 03, L = 04$, and $P = 06$. We want our message sender to encipher the message HELP in such a way that we, and only we, can decipher it. The sender knows we are using a 55-column array and that the public-key code is 7. He assigns $H = 2, E = 3, L = 4$, and $P = 6$. Next he must raise each number (2, 3, 4, and 6) to the seventh power and then find out which columns in our array the resulting numbers fall into. The sender sets up the following congruences:

$$H = 2; \text{ enciphered } H \equiv 2^7 \ (mod\ 55)$$
$$E = 3; \text{ enciphered } E \equiv 3^7 \ (mod\ 55)$$
$$L = 4; \text{ enciphered } L \equiv 4^7 \ (mod\ 55)$$
$$P = 6; \text{ enciphered } P \equiv 6^7 \ (mod\ 55)$$

Now how does he solve for the above congruences to find the corresponding columns? It's easy. Two raised to the seventh power is 128. Divide 128 by 55 and we get a remainder of 18. Hence, 2 enciphered becomes 18. In a similar manner we solve for 3, 4, and 6 to get 42, 49, and 41. Our original message has now been enciphered to be the four numbers: 18, 42, 49, and 41.

When we receive these four numbers we solve the following congruences to get the plaintext message out:

| | |
|---|---|
| First letter | $\equiv 18^{23} \equiv 2 \ (mod\ 55)$ |
| Second letter | $\equiv 42^{23} \equiv 3 \ (mod\ 55)$ |
| Third letter | $\equiv 49^{23} \equiv 4 \ (mod\ 55)$ |
| Fourth letter | $\equiv 41^{23} \equiv 6 \ (mod\ 55)$ |

Hence, by raising each of our enciphered numbers to the 23rd power and then finding the column each resulting number falls into, we get our original numbers back. We now look up how we assigned numbers to letters and discover the message reads "HELP."

After looking at these congruences, you may wonder how in the world we solve such a monster congruence containing 49 raised to the 23rd power. Using the laws of congruences it is relatively easy. We solve it by breaking it into a number of easier problems. To do this we use the laws of exponents which state: $A^{bc} = (A^b)^c$ and $A^{b+c} = A^b \cdot A^c$. Now we can write $X \equiv 49^{23}$ as

$$X \equiv (((49^2)^2)^2)^2 \cdot (49^2)^2 \cdot 49^2 \cdot 49 \ (mod \ 55)$$

To solve this monster we begin with the inside $49^2$ on the left and find the appropriate column by dividing by 55 and getting the remainder. Forty-nine squared is 2401. Dividing by 55 we get a remainder of 36. Hence, we substitute 36 for $49^2$ and have:

$$X \equiv (((36)^2)^2)^2 \cdot (49^2)^2 \cdot 49^2 \cdot 49 \ (mod \ 55)$$

Now we do the same with 36. Squaring and dividing by 55 gives a remainder of 31 and we have:

$$X \equiv ((31)^2)^2 \cdot (49^2)^2 \cdot 49^2 \cdot 49 \ (mod \ 55)$$

Continuing in the same fashion we get:

$$X \equiv (26)^2 \cdot (49^2)^2 \cdot 49^2 \cdot 49 \ (mod \ 55)$$

$$X \equiv 16 \cdot (49^2)^2 \cdot 49^2 \cdot 49 \ (mod \ 55)$$

$$X \equiv 16 \cdot (36)^2 \cdot 49^2 \cdot 49 \ (mod \ 55)$$

$$X \equiv 16 \cdot 31 \cdot 49^2 \cdot 49 \ (mod \ 55)$$

$$X \equiv 16 \cdot 31 \cdot 36 \cdot 49 \ (mod \ 55)$$

Now we can simply multiply the four numbers together and divide by 55 to get: $X \equiv 4 \ (mod \ 55)$ which is exactly the column number we wanted.

The above example was, of course, oversimplified. If you gave me the public key $E = 7$ and told me the array had 55 columns, I could quickly factor 55 to $5 \cdot 11$ and then compute $\phi(55)$. This would give me the basic equation: $7 \cdot D = 40k + 1$. Experimenting with different values for $k$ I would soon discover which one would yield a $D$ to decipher the message. How does one keep others from guessing the secret key $D$ in the system? Simple. You pick an array size $m$ that is very large and factors into only two large primes. The public knows $m$ but not the two primes $m$ factors into. Without knowing the two prime factors of $m$, no one can compute $\phi(m)$ since the formula for $\phi(m)$ depends on knowing the prime factors of $m$.

How large should $m$ be so it cannot be successfully factored? The developers of the public-key code use two large prime numbers which when multiplied together yield a 155-digit array size.[5] In order for someone to break this cipher and discover the decoding key, $D$, they would have to factor a 155-digit number to find the two primes. Once they knew the primes they could compute $\phi(m)$ and use this to solve for $D$.

How hard is it to factor a 155-digit number? The first factorization of a 100-digit number using a general purpose method was accomplished in October of 1988 by Arjen Lenstra and Mark Manasse.[6] The number was:

$$(11^{104} + 1)/(11^8 + 1) =$$

8675922231342839081221807709585070804897 ×

1084881048536374706129613998429729484098346115257905772116753

Initially, you might be tempted to think that a 155-digit number is only 55% larger than a 100-digit number. Be careful—a 101-digit number is ten times larger than a 100-digit number (just as 1000 is ten times larger than 100). A 155-digit number is $10^{55}$ larger than a 100-digit number.

It has been estimated by Carl Pomerance, an expert in number factorization, that at 1988 prices it would cost $10 million to factor a 144-digit number and $100 billion to factor a 200-digit number.[7] This hardly lends itself to easy assault by spies! We can make the cipher even harder to break. This is done by applying the cipher not to two-digit numbers representing single letters, but to groups of two-digit numbers. This ensures that an intruder cannot use recognition techniques from a language's common characteristics to break the code.

## CURRENT APPLICATIONS

We have reviewed the entire process for public-key codes and can recognize the importance of prime numbers in these ciphers. We must use an array number, $m$, which factors into only two large primes to insure that $\phi(m)$ is hard to find. If $m$ factors into any little primes, then it will be easy to factor completely. We must also recognize the importance of modern techniques for factoring large numbers. If factoring techniques become too successful, then larger array sizes must be used for public-key ciphers.

Companies and governments worldwide are now adopting the public-key cipher system. RSA Data Security, Inc. has grown to be the leading cryptographic marketing company in the nation. Almost every major hardware and software firm has obtained a license from RSA Data Security to develop products using the RSA encryption system. In addition to the many computer companies, RSA systems are used by the Federal Government and is bundled as part of Microsoft's new Windows 95 operating system.[8] One of RSA's products, MailSafe, provides secure communications for such organizations as Boeing, NATO, and Citibank.[9] European banks are adopting public-key systems. In the modern computer world where hackers steal data and plant viruses, the public-key cipher is proving to be a successful defense.

The RSA system is actually a combination of the DES cipher and their own public-key cipher. DES is used on the bulk of the message because it is faster than RSA, while the RSA system provides for public access to encrypting and signature verification. First, a DES key is randomly generated to mathematically scramble the digits of the message. The receiver has already selected two large primes which have been multiplied to yield a 155-digit number. This 155-digit number is expressed as a 512-bit binary number which is just the size computers like. The 155-digit number has yielded both the receiver's private key and the sender's public key. The sender also has a private key and public key determined by a different 155-digit number.

The sender creates a signature by using his own private key to encipher his name (and/or other identifying information). The

SENDER                              RECEIVER

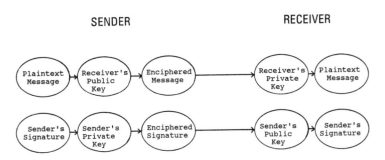

FIGURE 38.   Diagram of the public-key encryption system.

sender uses the receiver's public key to encipher the DES key which will unscramble the message. The sender then transmits the DES enciphered message, his enciphered "signature," and the enciphered DES key (Figure 38).

When the receiver gets the message, he first deciphers the DES key by using his own private key. This tells him how to decipher the message, which he does. To ensure that the message really came from the sender, the receiver uses the sender's public key to decipher his signature. Only the sender could have used his own private key to encipher his signature, hence the message must be authentic.

This all sounds complicated, but remember that computers do all the hard work. The entire process takes place in seconds or fractions of a second.

How do the RSA people know that the 155-digit numbers can't be factored and the enciphering system broken? Since 1990 RSA has been running a number factoring challenge. In this manner they hope to keep abreast of the latest factoring technology so they can assure their customers that the 155 numbers are large enough to remain unfactored, or they can design systems based on larger numbers. The contest offers cash prizes for factoring any of 41 numbers, each comprising two large primes, ranging from a relatively small 100-digit number up to a whopping 500-digit number. These numbers are designated as RSA-100, RSA-500, etc. A second list of easier numbers, called partition numbers, is also provided by RSA Data Security as a challenge for factoring. Anyone succeeding in factoring either an RSA number or a partition number can send their results to RSA Data Security. Their results will be verified and points awarded. Each second quarter of the year cash prizes are paid based on the points accumulated by various factoring teams or individuals. All the information regarding the factoring challenge sponsored by RSA Data Security can be found on the Internet's World Wide Web at http://www.rsa.com.

How successful have various factoring teams been in factoring the RSA challenge numbers? To answer this we must go back a number of years before the current challenge began. In 1977 the founders of RSA gave their first challenge to the mathematical community by providing RSA-129, a number with 129 digits. They claimed that to break such a large number and find its primes would take the fastest computers of 1977 running for millions of times the age of the universe. Some claim! This was all reported by Martin Gardner in his "Mathematical Games" column in *Scientific American*. However, in 1994, Derek Atkins and Arjen Lenstra led a factoring team that claimed the prize of $100, and not even a single age of the universe had passed.[10]

What had changed since 1977 when the challenge was first made? Two things: first, computers got faster, much faster. Second, new factoring algorithms were developed. Even so, the effort to crack RSA-129 took many months and the efforts of six hundred programmers and mathematicians. By checking the honor roll for factoring numbers maintained by RSA Data Security as part of their challenge, we see that RSA-100 was broken in 1991 by Arjen Lenstra and Mark Manasse for a $1000 cash prize offered by RSA Data Security. Then in 1993 RSA-110 was factored, yielding a prize of $5898. Of course, the time and cost of such efforts far exceed the cash prize offered.

To date only three RSA numbers have been factored, the original RSA-129, RSA-100, and RSA-110, out of the 42 RSA numbers. Each year the prize pot grows for the remaining numbers with the current kitty at $16,673. The partition numbers are much easier to factor, and as of June 1995, over 1200 had been cracked, with cash prizes exceeding $22,000. By the time you read this, many other numbers may have been factored. Therefore, assuming you wish to "turn your hand" at factoring a large number I give you RSA-500, the largest RSA number in the challenge:

18971941337486266563305347433172025272371835591953428303184
58112306245045887076876059432123476257664274945547644419515
42758674320565931725466994660498241973016010381252152852854
68803151640161162396312837062979326593940508107758169447866
04172141102464103804027870110980866421480002556045468762513
77453934182215494821277335671735153472656328448001134940920
64424384401989109086032526788147850601132077287172819942445
11323201949222955423789860663107489107472242561739680319169
24381467623571293429229997441361

This 500-digit number should keep you busy for a few nights.[11]

## SINK THE CLIPPER!

With the phenomenal growth in use of public-key encryption methods, such as the RSA system, we might be led to believe that public-key codes will soon become the national, or even world standard. However, the U.S. Government has been fighting to avoid such a situation. The National Security Agency of the U.S. Government is responsible for breaking the codes of foreign governments—they are our secret spy agency. But the NSA knows that it cannot break the public-key codes. Hence, other governments and criminals could adopt public-key codes with the assurance that our government cannot eavesdrop. No wonder the NSA is upset about the RSA success story.

The government has not been idle, and is now working on a competing system to public-key encryption. In 1987, Congress authorized the National Institute of Standards and Technology (NIST) to develop an acceptable encryption system that would satisfy the needs of user privacy, yet allow law enforcement agencies and the NSA to decipher transmitted messages. This effort became the Capstone project. Under Capstone, a computer chip, called a Clipper chip, would be manufactured and installed in computers that interface with the U.S. Government. The Clipper chip would encode and decode messages for users. However, if the government demonstrated a need for breaking a data transmission they could acquire the secret user keys from two escrow (holding) agencies that would each possess a different part of the key. Hence, the government could, under special conditions, decipher any messages from Capstone.

Objections to the Capstone system are twofold.[12] Critics claim that the very idea of our government eavesdropping on computer communications is contrary to our hallowed ideas of personal freedom. They also say that too many features of the Capstone system are secret, and therefore, the system is untested, and may not provide a truly secure encryption service.

How these issues will be resolved is difficult to say. Certainly, RSA Data Security, Inc. has an overriding interest in seeing the public-key system adopted and not the Capstone system. They have, in fact, launched a public relations campaign to "Sink the Clipper!" How any individual feels about the outcome of this struggle depends on how much weight is given to our government's need to monitor foreign governments and criminal elements, as opposed to our rights of privacy.

## CHAPTER TEN

# THE REMARKABLE RAMANUJAN

*The millions are awake enough for physical labor; but*

*only one in a million is awake enough for effective*

*intellectual exertion, only one in a hundred millions*

*to a poetic or divine life. To be awake is to be alive.*

HENRY DAVID THOREAU

*WALDEN*[1]

Peter Ratener, mathematics professor at Bellevue Community College in Washington State, once told me that while interviewing prospective candidates for staff positions within the department, he liked to ask them to name their favorite 20th century mathematician. In most cases, he was disappointed to report, the candidates simply stared at him, unable to think of a single modern mathematician. If we take a moment to consider this reaction, we will see just how astounding it is. How many graduates with a master's degree in American Literature would not be able to name a single 20th century author? How many psychology graduates would be unable to mention Freud or Jung? Yet, incredible as it is, mathematics graduates from American universities are nearly blind to the identities of modern mathematicians. This situation has come about because we have successfully stripped all cultural identifiers from our mathematical knowledge.

The mathematicians we may occasionally hear of are generally from the remote past: Pythagoras, Euclid, and Archimedes. Even the more recent ones worked hundreds of years ago: Newton, Galileo, and Descartes. No one alive today ever knew these men, and when we describe them as brilliant, creative, and genius, we have no real sense that we know them. In fact, many of these famous personalities are known not as mathematicians, but for other works: Newton for discovering gravity, Galileo for using the telescope, and Descartes for his philosophy. We are about to change all this, for we are going to introduce two great 20th century mathematicians: Srinivasa Ramanujan and Godfrey Harold Hardy. One was certainly a genius, and the other a brilliant mathematical talent. Through them we can get a vicarious sense of greatness in a time not too distant from today.

## OF HUMBLE BIRTH

Srinivasa Ramanujan Iyengar (Figure 39) was born on Thursday, December 22, 1887, on Teppukulam Street in the city of Erode, Southern India.[2] While his family was of the Brahmin caste, they were poor; his father worked as an accountant for a cloth merchant. His first name, Srinivasa, was that of his father, while his last name, Iyengar, was his caste name. Therefore, the name he went by was simply Ramanujan or sometimes S. Ramanujan. He was the first child born to his mother, Komalatammal. She would become a doting and protective mother to her eldest son.

A short, pudgy boy who was slow to learn to speak, Ramanujan early showed a strong inclination toward mathematics. In elementary school he won numerous awards for his calculating skills. Upon graduation in 1904, he won a scholarship to college.

The India of 1904 was an unlikely place to nurture unusual mathematical talents. The British had ruled India for almost 200 years, and the educational system they fostered was meant to train loyal civil servants who would, in turn, help manage a large, populous, and technologically backward country. The system was not designed to identify creative and imaginative leaders.

Sometime between the 15th and 12th centuries B.C., Aryans invaded India from the north and established a caste system. Since then, the privileged Brahmin caste produced the majority of India's

FIGURE 39. Srinivasa Ramanujan Iyengar, 1887–1920.

priests, scholars, and gurus. Yet, Ramanujan's membership in India's highest caste did not guarantee that he would find success. His family was not rich and had to struggle to send him to Government College in Kumbakonam, even with the scholarship. Therefore, in August 1905 when he failed his examinations and lost the scholarship, his chance for success looked dim. He failed through neglect of his studies—except mathematics. In fact, he seemed to spend all his time scribbling strange symbols on his slate board, using his elbow as an eraser, or writing in his personal notebooks—notebooks crammed with more mathematical symbols. Yet, his mother indulged him, and in 1906 bundled him off to Pachaiyappa College in Madras, the third largest city in India.

Evidently, Ramanujan hadn't learned from his first experience. In 1907 he failed his examinations again. Almost 20, he had flunked out of two colleges, had no job or degree, and no good prospects for making something of himself. For the next few years he did little but hang around and live within the private world of his note-books. Was he destined to simply fade into obscurity along with millions of other bright young men who seem to lose their way while growing into manhood? His mother wasn't about to let that happen. She decided it was time he married. As was the custom in India, Komalatammal arranged for her son's marriage. The bride was Srimathi Janaki, a ten-year-old girl of the same caste in a neighboring town. On July 14, 1909, Ramanujan and Janaki took their vows, even though she would not actually live with Ramanu-jan and his family for several more years.[3] However, things did not suddenly turn around for the young man. For the next three years he held no serious job, spending most of his time on his mathemat-ics while friends and associates indulged him by providing a small stipend to live on. During this period Ramanujan showed his notebooks to any interested party in the hopes of drawing some attention to his work. The problem with this approach was that no one in Southern India at this time was sufficiently trained to appreciate Ramanujan's talents and, at the same time, promote his future.

However, he did succeed in publishing his first mathematics paper in the *Journal of the Indian Mathematical Society* in 1911. In his article he proposed two problems and solicited their solutions. One of these problems involved an infinite radical, something we are familiar with. The problem was to find the value of:

$$\sqrt{1 + 2\sqrt{1 + 3\sqrt{1 + 4\sqrt{1 + \ldots}}}}$$

This is a simple problem to understand, but somewhat difficult to solve. What is remarkable is that Ramanujan, without benefit of a formal higher education in mathematics, understood the problem and, in fact, knew the answer. Could the *Journal*'s mathematician readers solve the puzzle? Evidently not, for six months and three more issues passed without any of its readers offering a solution. Finally, Ramanujan gave the answer himself—it was simply 3. But the amazing thing is not just that Ramanujan knew both the problem and the answer, but how he knew of them, for on page 105 of his first notebook was a general solution to this kind of problem.[4]

$$x + n + a = \sqrt{ax + (n + a)^2 + x\sqrt{a(x + n) + (n + a)^2 + (x + n)\sqrt{\cdots}}}$$

The problem he posed in the *Journal* was a special case of this equation (when we let $x = 2$, $n = 1$, and $a = 0$). What other marvelous wonders were hidden in those notebooks?

Suddenly, things began to happen. In 1912 Ramanujan secured a job as a clerk in the accounts section of the Madras Port Trust which paid 30 rupees a month (£20 per year). The Chairman of the Port Trust was an English engineer, Sir Francis Spring, and the Trust's manager was an Indian mathematician, S.N. Aiyar. Both men encouraged Ramanujan to contact English mathematicians in the hope that they would assist him in getting his work published.[5]

During the year he wrote letters to two well-known English mathematicians, Henry F. Baker and E.W. Hobson, both fellows of the Royal Society. With his letters, Ramanujan included samples of his work from his notebooks, asking for their assistance. Both mathematicians turned him down.

Then in 1913 he wrote a ten-page letter to the famous mathematician, G. H. Hardy (Figure 40), an author of three books and over 100 articles, a Fellow at Trinity College, Cambridge, and also a Fellow of the English Royal Society. This seed found fertile ground.

Hardy, upon first reading Ramanujan's letter, wondered if maybe this wasn't some sort of joke instigated by his playful comrades at Trinity. An uneducated Indian boy doing advanced mathematics! He was amazed and intrigued by the theorems that Ramanujan had included in the letter. While Hardy was a world-class mathematician, there was one other at Cambridge who would also be qualified to understand the formulas in the letter—his friend and collaborator, John Littlewood (1885–1977), one of the world's great number theorists (Figure 41).

FIGURE 40.   Godfrey Harold Hardy, 1877–1947.

That evening Hardy took Ramanujan's letter to the chess room
above the commons room at Trinity College. He and Littlewood
pored over the theorems for 2 1/2 hours, and when they finished
they could reach only one conclusion. This Ramanujan, whoever
he was, was a first-class, world mathematician. Somehow they
would have to get Ramanujan to Cambridge. One of Hardy's
associates, Eric Neville, was on his way to India to deliver a series
of lectures at the University of Madras. Hardy instructed Neville
to find Ramanujan and convince him to come to England.

FIGURE 41. John Littlewood, 1885-1977.

Neville did as Hardy asked, but a problem developed. The Brahmin caste believes travel to a foreign country is an unclean act and forbids it. Ramanujan would be shunned by friends and family if he traveled abroad. Yet, the English mathematicians wanted Ramanujan at Cambridge so they could collaborate with him in person; the mails would be too slow, awkward, and impersonal. The solution came from Ramanujan, himself. He was a devotedly religious young man, adhering strictly to his Hindu beliefs. Fortunately, he had a vision from his personal Hindu god, the Goddess Namakkal, giving him permission to go. Once this barrier was surmounted, Neville and Hardy scrambled to secure the funds needed to provide Ramanujan an income and pay his passage. Finally, on April 14, 1914, when he was 26, Ramanujan reached London. He was now in a strange land, without wife, family, or friends. Yet, he had come to work with some of the world's best mathematicians—his life would never be the same.

## A STRANGE JUXTAPOSITION

For the next five years, Ramanujan was associated with Hardy of Trinity College at Cambridge University. Their collaboration represents the efforts of two great talents, and two personal backgrounds that could not have been more different. Hardy, while not of the aristocratic class in England, came from middle-class parents who were both educators. Not from royalty, yet educated in one of England's finest schools, he always showed a somewhat proletarian resistance to aristocratic backgrounds. He was an outspoken atheist and a pacifist. He belonged to a famous "secret" society at Trinity College known as the Apostles (established in 1820), which was chock full of famous English thinkers, including Bertrand Russell, James Clerk Maxwell, Alfred Tennyson, and Alfred North Whitehead. Numerous members were known to be homosexual at a time when "coming out of the closet" was not fashionable or advised. Whether Hardy, himself, was a homosexual is not known, but he had no meaningful relationships with women during his life except with his mother and sister, Gertrude.

Ramanujan was at the other extreme. He had not received a strong education growing up in South India. Indeed, he had never even received a two-year associate's degree from college. Unlike Hardy, he was a deeply religious man, holding firmly to his Hindu beliefs. All the time he was in England he remained a vegetarian, even though his favorite foods generally were not available, especially during the war years. He was married, but not a father. These two men, Hardy of England, and Ramanujan of India, had little in common—except mathematics. Because of this, they were collaborators more than friends. Hardy confessed in later years that he knew few details of Ramanujan's personal life. While the two men worked together, they did not socialize or become "buddies." However, there can be no doubt that they respected and cared for each other.

Sometime during the war years, Ramanujan contracted a mysterious illness, and was sent to several sanatoriums. At times he would improve, but then lapse back into poor health. Because of the war, it was unsafe to return by ship to India. Finally, in 1918

World War I came to an end. In poor health and homesick, Ramanujan left England and arrived in India in March 1919. Despite being back in the care of his mother and his wife, his illness could not be rolled back, and he worsened. During this time he continued to work on his beloved mathematics. On April 26, 1920, Ramanujan died at the age of 32 while his wife, Janaki, attended to him. Some have guessed that the illness was tuberculosis, but this has never been established.[6]

Hardy, upon hearing the news, realized that the one person who would prove to exert the profoundest influence on his life had passed away. Hardy continued with his mathematics for another 27 years. He died in 1947 at the age of 70.

## THE RAMANUJAN LEGACY

Just what did Ramanujan leave? While he completed significant work in number theory with Hardy while in England, becoming both a Trinity Fellow and a Fellow of the Royal Society, it is his personal notebooks that are especially intriguing. It has been estimated that they contain between three and four thousand theorems, as much as two-thirds being new to mathematics, the balance representing independent rediscoveries of other mathematicians' work.[7] Much of Ramanujan's study was in number theory, the very material we have been covering ourselves. Ramanujan was no stickler for formal mathematical proofs. Once he had discovered a relationship and satisfied himself it was true, he went on with the next problem. Hardy, on the other hand, stood squarely in the camp that insisted on rigorous demonstrations. Only with such proofs could we know the truths of mathematics were certain and could then be used in future proofs. The two talents complemented each other perfectly. Ramanujan, an undisputed genius, could somehow take that deep look into the inner heart of mathematics and pull out a beautiful equation. Hardy possessed the strength of heart and mind to insist on proving the equations so they became part of humankind's vast storehouse of knowledge.

When we comprehend some of Ramanujan's equations, we realize that he was a true artist, expressing deep and beautiful mathematical truth in familiar symbols. To appreciate his artistry, all we have to do is look at a few of the equations he included in his letter to Hardy. Some of the following equations contain concepts that are beyond the scope of our work, yet in them we see a wonderful balance and symmetry.[8]

$$1 - 5\left(\frac{1}{2}\right)^3 + 9\left(\frac{1\cdot3}{2\cdot4}\right)^3 - 13\left(\frac{1\cdot3\cdot5}{2\cdot4\cdot6}\right)^3 + \cdots = \frac{2}{\pi}$$

Here we have a beautiful infinite series summing to 2 divided by $\pi$. We can't escape delighting in its elegant form, with the coefficients, 1, 5, 9, and 13 all differing by 4, and their signs alternating. Within the parentheses we find numerators as products of successive odd numbers and the denominators as products of successive even numbers.

The next equation involves both an integral from calculus and an infinite continued fraction.

$$\int_0^a e^{-x^2} dx = \frac{1}{2}\pi^{1/2} - \cfrac{e^{-a^2}}{2a + \cfrac{1}{a + \cfrac{2}{2a + \cfrac{3}{a + \cfrac{4}{2a + \ldots}}}}}$$

Even though we cannot fully appreciate this equation without calculus, we are still moved by its intricate pattern. An additional bonus is that it relates Euler's $e$ to $\pi$.

This next Ramanujan equation contains a hidden surprise. See if you can identify it.

$$\cfrac{1}{1 + \cfrac{e^{-2\pi}}{1 + \cfrac{e^{-4\pi}}{1 + \ldots}}} = \left[\sqrt{\left(\frac{5 + \sqrt5}{2}\right)} - \frac{\sqrt5 + 1}{2}\right] \cdot e^{\frac{2}{5}\pi}$$

Can you see what's hidden in the right side of the equation? It's $(\sqrt5 + 1)/2$—the Golden Mean! We also find it under the square root on the right since:

$$\frac{5 + \sqrt{5}}{2} = \frac{4 + \sqrt{5} + 1}{2} = 2 + \frac{\sqrt{5} + 1}{2}$$

Therefore, if we substitute $\phi$ for $(\sqrt{5} + 1)/2$ we get the following simplified equation:

$$\cfrac{1}{1 + \cfrac{e^{-2\pi}}{1 + \cfrac{e^{-4\pi}}{1 + \ldots}}} = (\sqrt{2 + \phi} - \phi) \cdot e^{\frac{2}{5}\pi}$$

Now we have an expression including not only an infinite continuing fraction, $e$, and $\pi$, but the Golden Mean, too! How did Ramanujan come up with this? Did he know the definition of the Golden Mean, and if so, why didn't he use a single symbol for $(\sqrt{5} + 1)/2$? How did he discover this relationship with so little formal schooling? Something magical must have been going on inside his head for him to have made such intuitive leaps. Ramanujan, himself, credited his discoveries to his family deity, the Hindu goddess Namakkal. Is it any wonder that Hardy was astounded when he read Ramanujan's letter containing these beautiful equations?

Ramanujan's ability to peer deeply into the inner chambers of mathematics was probably not unique, for we have stories of other great mathematicians who seem to have had similar talent. Among them we would probably want to include Archimedes, Newton, Euler, and Gauss. In all likelihood, Pythagoras and Euclid were also of this bent. And while there are many truly brilliant mathematicians, there are relatively few with this gift of the golden eye to see into the recesses to which the rest of us are blind. It is sad that Ramanujan died so young, for he probably had much more to give to humanity. And his example should serve as a warning to us. It is not guaranteed that a great mind will automatically find the nurturing support required for success. If Ramanujan had not been a Brahmin, or his mother had not been as patient, or if Hardy had ignored his letter, then Ramanujan might have slipped into obscurity, and his wonderful notebooks and equations would have been lost forever. How many geniuses have never been given the chance to share their gifts? Surely they are too few that humankind can afford to lose any.

## THE HARDY LEGACY

We cannot leave Ramanujan's story without remarking on the impact of Hardy's work. Hardy was a world-class mathematician, a captivating lecturer, and a fascinating personality. Several of his books have become classics, and I have two of them on my own shelves.[9] He and his friend, John Littlewood, established the field of analytical number theory. Yet, it is not his published work that makes us focus on him here. What we are concerned with is Hardy's attitude toward mathematics, one which he helped popularize among other mathematicians.

Hardly a professional mathematician alive today has not read Hardy's *A Mathematician's Apology*, a short essay outlining Hardy's aesthetic ideas regarding mathematics. His attitude can be summed up simply: pure mathematics is good and beautiful; applied mathematics is somehow ugly or debased. Pure mathematics is, of course, mathematics studied for its own sake, without any concern for whether it solves any problems in the physical universe. In the *Apology*, Hardy says:

> I will say only that if a chess problem is, in the crude sense, 'useless,' then that is equally true of most of the best mathematics; that very little of mathematics is useful practically, and that that little is comparatively dull.[10]

Possibly Hardy's disdain for applied mathematics can, in part, be traced to his pacifism. In a letter written just after World War I we find:

> I must leave it to the engineers and the chemists to expound, with justly prophetic fervor, the benefits conferred on civilization by gas-engines, oil, and explosives. If I could attain every scientific ambition of my life, the frontiers of the Empire would not be advanced . . .[11]

If applying mathematics could result in the construction of modern weapons, weapons which result in so much death and destruction, then certainly this was cause to reject such mathematics. Again in the *Apology* he says,

> I have never done anything 'useful.' No discovery of mine has made, or is likely to make, directly or indirectly, for good or ill, the least difference to the amenity of the world.[12]

Hardy believed that beauty was the dominant characteristic of a mathematical theorem, and that mathematical beauty was dependent upon the theorem's seriousness. This seriousness was measured by how much the ideas within the theorem were connected to other mathematical ideas.

> The mathematician's patterns, like the painter's or the poet's, must be beautiful.... Beauty is the first test: there is no permanent place in the world for ugly mathematics.... The best mathematics is *serious* as well as beautiful.... The 'seriousness' of a mathematical theorem lies, not in its practical consequences, which are usually negligible, but in the *significance* of the mathematical ideas which it connects.[13]

In summary, Hardy believed a mathematical idea is good because it is beautiful, beautiful because it is serious, and serious because it is connected to many other mathematical ideas. As stated, Hardy's claim that beauty is central to the enjoyment of mathematics is fervently believed by the majority of all who are enthralled by mathematics. In this he seems to have captured the essence of our love for this subject matter. Jerry King, in *The Art of Mathematics*, points out, "Mathematicians know beauty when they see it for that is what motivates them to do mathematics in the first place."[14]

However, Hardy went beyond the claim that mathematics is beautiful to also insist that applications of mathematical ideas to the physical world demean those ideas, and that the beauty of a mathematical idea is not connected to its usefulness—its usefulness detracts from its beauty.

Hardy's idea that pure (good) mathematics should be devoid of meaningful applications has been adopted by many mathematicians at our universities. Unfortunately, this idea has caused some mathematicians to become elitists, casting disdain on all other branches of knowledge. This, in turn, has tended to alienate mathematicians from the rest of the academic community. Most elementary and secondary teachers we send out of our universities are not professional mathematicians, and they feel this alienation between themselves and what they see as snobbish old men barricaded in the ivory towers of academia. These same teachers, who feel alienated from higher mathematics, are asked by us to teach our children the foundations of mathematics. Do you imagine they embrace the task with enthusiasm?

Hardy's legacy, that applied mathematics is somehow spoiled, has caused much harm to the general study of mathematics, in addition to creating an artificial separation between pure and applied mathematicians. The notion that finding an application for a mathematical idea somehow demeans that idea is silly, as a little thought will demonstrate. This idea of Hardy's is, in fact, new to this century, for the great preponderance of past mathematicians contributed to both pure and applied fields. Those three names most often mentioned as the greatest mathematicians to have ever lived are Archimedes, Newton, and Gauss, each of whom chose to explore and contribute heavily to numerous fields in addition to pure mathematics. Archimedes (287–212 B.C.) helped defend the ancient city of Syracuse against the Roman army during the Second Punic War. He designed and supervised the construction of great war machines to repel the attackers. Newton (1642–1727) not only coinvented calculus, but discovered the law of gravitation. For almost his entire career, Gauss (1777–1855) was director of the astronomical observatory at Göttingen, making significant contributions in the fields of astronomy and physics.

The truth behind beauty is difficult to track down, but we do understand one characteristic of beauty as illustrated by the old cliche, "Beauty is in the eye of the beholder." To be an engineer, and build a marvelous machine, and to see the beauty of its operation is as valid an experience of beauty as a mathematician's absorption in a wondrous theorem. One is not "more" beautiful than the other. To see a space shuttle standing on the launch pad, the vented gases escaping, and witness the thunderous blast-off as it climbs heavenward on a pillar of flame—this is beauty. Yet it is a prime example of applied mathematics.

It is past time that elitist ideas of mathematics be set aside for an open-minded approach. For most of this century the general population has viewed mathematics as some strange game of symbols played by boring old men in dusty rooms, hidden away on our university campuses. The truth about mathematics is far different. The work of pure mathematicians is constantly being used by others—astronomers, computer scientists, engineers—to forge ahead with exciting breakthroughs and innovations that are changing and improving the world we live in.

The approaching century suggests a new beginning, and we can use this as an excuse to revitalize the general perception of mathematics. We must find teachers who truly love mathematics to teach our young. This places personal responsibility on each person who is enthralled by mathematics to contribute to improving primary and secondary mathematics education. We can have a general population that likes and respects mathematics, both in its elegant beauty and its usefulness.

## CHAPTER ELEVEN

# RAMANUJAN'S EQUATIONS

*But he who has been earnest in the love of knowledge*

*and of true wisdom, and has exercised his intellect*

*more than any other part of him, must have thoughts*

*immortal and divine, if he attain truth, and in so far*

*as human nature is capable of sharing in immortality,*

*he must altogether be immortal; . . .*

PLATO

*TIMEUS*[1]

## LEARNING TO LOVE EQUATIONS

We are now going to look at more of Ramanujan's equations. I know that for some of you, the prospect of facing additional equations causes your heart to palpitate, and your palms to sweat. "Why," you say, "does he have to use more of those darn equations? Why can't he just say it in ordinary words?"

We recognize that a broad segment of the population does not like equations. Yet, to fully understand what mathematicians do and why they do it, we must come to terms with equations; in fact, we must come to like, nay, *love* equations. All true mathematicians love equations, for they are to the mathematician what a masterpiece is to the art connoisseur. Mathematics is about relationships, and equations are the symbolic language of relationships. While it

216

is true we could say the same thing in plain English, the result in most cases would be an entire paragraph whose meaning was difficult to decipher. In fact, using ordinary language is how mathematical relationships were expressed before the development of symbolic mathematics. A quick look at one of Euclid's propositions from 300 B.C. illustrates the point.

> If a first magnitude be the same multiple of a second that a third is of a fourth, and a fifth also be the same multiple of the second that a sixth is of the fourth, the sum of the first and fifth will also be the same multiple of the second that the sum of the third and sixth is of the fourth.[2]

To be fair to Euclid, we must point out that most of his definitions and propositions are easier to understand than the one quoted, and that symbolic notation was not available to him. However, what we are dramatizing here is that mathematics had to abandon ordinary language and develop a symbolic language in order to clearly state complex ideas. Therefore, we should not look at mathematical symbolism as the enemy to be avoided, but as a great tool which we can use to express deep and beautiful ideas.

Now that I have you convinced that mathematical symbols and their associated equations are absolutely necessary, how does that fact help you to appreciate them? We can approach equations much as a lover of fine art studies a great painting. Once we understand what distinguishes a masterpiece from ordinary artistic flotsam, we quickly recognize its beauty. To do this with equations we must recognize several characteristics of equations, and how these characteristics impact the equation's aesthetic appeal.

The first job is to quickly scan the equation to determine if it contains any symbols with which we are unfamiliar. If so, we must check some reference or textbook to learn their meaning. Authors of higher mathematics texts frequently place sections in the beginnings of their books listing those symbolic conventions the author is using. Once we are satisfied that we understand the basic meaning of each symbol, we are ready to take the next step.

Now we wish to "step back" from the equation and ask in general terms, what is being equated on each side of the equal sign.

This will help us to reach the first level of understanding of the equation's meaning—that is, the deep truth which the relationship is stating. For example, a simple identity could be the equation: 1 + 1 = 2. On the left of the equal sign we have two terms and on the right, one term, all of which are integers. While the equation is true (by definition), it is not terribly interesting because we are too familiar with the idea expressed.

What about Euler's beautiful equation:

$$e^{\pi\sqrt{-1}} + 1 = 0$$

Here again we have only two terms on the left, and one on the right. However, these individual terms are what make the equation so beautiful. The equation relates the constants $e$ and $\pi$, the square root of $-1$, the number 1, and the number 0. Hence, it is the identity and interconnectedness of terms found in the equation that gives the equation its richness. Therefore, we must be on the lookout for interesting terms present in our equations, and how the equation relates these terms.

Next, a beautiful equation is frequently aesthetically pleasing to the eye by its form and symmetry. As an example, the sum of the reciprocals of all natural numbers squared equals $\pi^2/6$, an identity discovered by Euler.

$$\sum_{n=1}^{\infty} \frac{1}{n^2} = \frac{1}{1^2} + \frac{1}{2^2} + \frac{1}{3^2} + \frac{1}{4^2} + \frac{1}{5^2} + \ldots = \frac{\pi^2}{6}$$

In this relationship we immediately notice several things. On the left is an infinite series of terms (expressed two ways). On the right is a constant involving $\pi$. The fact that an infinite sum of terms even has a finite limit is interesting, let alone the fact that the limit involves $\pi$. We have actually expressed the infinite series in two ways. On the far left is a compressed form, and in the middle is an expanded form. While most books on higher mathematics favor the compressed form, such a form frequently hides the series' beautiful symmetry. Therefore, it will be our custom to generally give both the compressed form and the expanded form of infinite series, continued fractions, and continued radicals.

Looking at the expanded form of the above infinite series, we recognize at once the marvelous symmetry within, a kind of symbolic harmony. Every numerator is a 1, every exponent in the denominators is a 2. Each $n$ in the denominator increases by exactly 1 from its predecessor. The harmony in this infinite series can almost be tasted, as a delectable morsel of Swiss chocolate.

## RAMANUJAN'S ARTISTRY

We are now ready to enjoy more of Ramanujan's work. I have selected only those equations involving mathematical concepts that we have already covered. Many lovely relationships will not be reviewed, including those involving integrals (calculus), the gamma function, Bernoulli numbers, and Euler numbers. While Ramanujan was not trained in the mathematics of complex numbers (complex analysis), many of his theorems have been proven true for not only real numbers but also for complex numbers. We will ignore complex numbers in our review.

Ramanujan was an expert on infinite series, continued fractions, and other exotic mathematical creatures. Frequently he would show the general form of a relation, which he called an entry, followed by several corollaries, or specific examples of the general equation. Most of our examples are drawn from the corollaries. In those cases where we give the general form, we will also provide several examples. Our first equation involves an infinite series which adds to zero.[3]

$$\sum_{k=0}^{\infty} = (-1)^k \frac{(2k+1)^3 + (2k+1)^2}{k!} = \frac{1^3 + 1^2}{0!} - \frac{3^3 + 3^2}{1!} + \frac{5^3 + 5^2}{2!} - \ldots = 0$$

Checking for any terms we might be unfamiliar with, we see the factorial sign (!) in the denominators of our fractions. Remember that $n!$ stands for all the integers from 1 through $n$ multiplied together. Hence, 5! is simply $1 \cdot 2 \cdot 3 \cdot 4 \cdot 5 = 120$. By definition, $0! = 1$. Therefore, the first term of the expanded infinite series $(1^3 + 1^2)/0!$ is equal to $(1 + 1)/1$ or simply 2. Take a moment to savor the pattern found in the expanded infinite series. The terms in the numerators are the successive odd integers. The exponents are always 3 and 2.

The denominators are the factorials of successive integers beginning with zero. Just looking at the expanded form we know at once what the next term is going to be: $(7^3 + 7^2)/3!$ and the term after that will have nines in the numerator and $4!$ in the denominator. Notice also that the terms in the series are alternately added and subtracted. We get this from the term $(-1)^k$ found in the compressed form, which does nothing but alternate the signs of successive terms. If we take the trouble to actually calculate the terms in the expanded form of the infinite series, we see an interesting phenomena. The first three terms are $+2$, $-36$, and $+75$. It would appear the successive terms are growing large, but with the very next term we see them begin to diminish, e.g., $-65.33$, $+33.75$, $-12.1$, etc. If we add consecutive terms together, we notice that the partial sums to the infinite series get closer to zero, or:

| | |
|---|---|
| 2 | $=$ 2.00 |
| $2 - 36$ | $= -34.00$ |
| $2 - 36 + 75$ | $=$ 41.00 |
| $2 - 36 + 75 - 65.33$ | $= -24.33$ |
| $2 - 36 + 75 - 65.33 + 33.75$ | $=$ 9.42 |
| $2 - 36 + 75 - 65.33 + 33.75 - 12.1$ | $= -2.68$ |
| $2 - 36 + 75 - 65.33 + 33.75 - 12.1 + 3.29$ | $=$ 0.61 |
| $2 - 36 + 75 - 65.33 + 33.75 - 12.1 + 3.29 - 0.71$ | $= -0.10$ |

Of course, the sum of the infinity of terms of the series will be equal to exactly zero.

Our next equation involves both an infinite series and $\pi$.

$$\sum_{k=1}^{\infty} \frac{1}{k^3(k+1)^3} = \frac{1}{1^3 \cdot 2^3} + \frac{1}{2^3 \cdot 3^3} + \frac{1}{3^3 \cdot 4^3} + \frac{1}{4^3 \cdot 5^3} + \cdots = 10 - \pi^2$$

Here, again, we find the beautiful symmetry that is characteristic of Ramanujan's work: All ones in the numerators, while the denominators are products of the cubes of successive integers. This infinite series sums to exactly $10 - \pi^2$ or approximately $0.130395$.

Next we have a series whose sum is the natural logarithm of 2. Remember that natural logarithms are simply the exponents on the constant $e$. The natural logarithm of 2 is that power we must raise $e$ to in order for it to equal 2: $e^{\ln 2} = 2$. While $\ln 2$ is a transcendental number, and cannot be written exactly in decimal form, it is equal to approximately 0.693147.

$$\frac{1}{2} + \sum_{k=1}^{\infty} \frac{1}{(2k)^3 - 2k} = \frac{1}{2} + \frac{1}{2^3 - 2} + \frac{1}{4^3 - 4} + \frac{1}{6^3 - 6} + \frac{1}{8^3 - 8} + \cdots = \ln 2$$

Once again we find the pattern which allows us to immediately construct the next term in the expanded version or $1/(10^3 - 10)$.

The next equation also involves a natural logarithm, but multiplies the infinite series by the constant 2.

$$1 + 2 \cdot \sum_{k=1}^{\infty} \frac{1}{(3k)^3 - 3k}$$

$$= 1 + 2 \left[ \frac{1}{3^3 - 3} + \frac{1}{6^3 - 6} + \frac{1}{9^3 - 9} + \frac{1}{12^3 - 12} + \cdots \right] = \ln 3$$

Another way of thinking of the above equation is to realize that if we computed the left side of the equation and used that as the exponent on the constant $e$ the result would be equal to 3.

This next equation relates both a logarithm and $\pi$ to an infinite series.

$$\sum_{k=1}^{\infty} \frac{\left(\frac{1}{2}\right)^k}{k^2} = \frac{\frac{1}{2}}{1^2} + \frac{\left(\frac{1}{2}\right)^2}{2^2} + \frac{\left(\frac{1}{2}\right)^3}{3^2} + \frac{\left(\frac{1}{2}\right)^4}{4^2} + \cdots = \frac{\pi^2}{12} - \left(\frac{1}{2}\right)(\ln 2)^2$$

Notice as we calculate the values of successive terms in the infinite series, the numerators grow small while the denominators grow large. This causes the series to converge rather quickly to the limit on the right.

Our next example includes not only $\pi$ and a natural logarithm, but an additional surprise. See if you can find it.

$$\sum_{k=0}^{\infty} \frac{(-1)^k (k!)^2}{(2k)!(2k+1)^2}$$

$$= 1 - \frac{1}{2!\cdot 3^2} + \frac{(2!)^2}{4!\cdot 5^2} - \frac{(3!)^2}{6!\cdot 7^2} + \frac{(4!)^2}{8!\cdot 9^2} + \ldots = \frac{\pi^2}{6} - 3\cdot ln\left(\frac{\sqrt{5}+1}{2}\right)^2$$

Do you recognize the term on the right? It is our Golden Mean or $(\sqrt{5}+1)/2$. Hence, we have an infinite series related to $\pi$, a natural logarithm, and the Golden Mean! As with all the other equations, we have the nice Ramanujan symmetry.

Now we have an infinite series involving one unknown variable, $x$. This is a general form of an equation into which we can substitute specific values for $x$ and get a new formula. The only restriction on $x$ is that it must be greater than $1/2$.

$$1 - \frac{x-1}{x+1} + \frac{(x-1)(x-2)}{(x+1)(x+2)} - \frac{(x-1)(x-2)(x-3)}{(x+1)(x+2)(x+3)} + \ldots = \frac{x}{2x-1}$$

We see at once what the next term is going to be: It will include $(x-4)$ in the numerator and $(x+4)$ in the denominator. If we substitute a value for $x$ that is a positive integer, one of the numerators in one of the terms will become zero, causing all the numerators (and therefore terms) that follow to be zero, leaving us with a finite series. For example, if we let $x = 2$ we get:

$$1 - \frac{2-1}{2+1} + \frac{(2-1)(2-2)}{(2+1)(2+2)} - \frac{(2-1)(2-2)(2-3)}{(2+1)(2+2)(2+3)} + \ldots = \frac{2}{4-1}$$

Simplifyingwehave:

$$1 - \frac{1}{3} + \frac{0}{12} - \frac{0}{60} + \ldots = \frac{2}{3}$$

To make an interesting substitution we must choose an $x$ that is not a positive integer. Suppose we let $x$ equal 2.5. Then the above equation becomes:

$$1 - \frac{2.5-1}{2.5+1} + \frac{(2.5-1)(2.5-2)}{(2.5+1)(2.5+2)} - \frac{(2.5-1)(2.5-2)(2.5-3)}{(2.5+1)(2.5+2)(2.5+3)} + \ldots$$

$$= \frac{2.5}{2(2.5)-1}$$

Combining terms and simplifying we get the infinite series:

$$1 - 0.4286 + 0.0476 + 0.0043 + 0.0010 + \ldots = 0.625$$

We could, of course, make any number of substitutions for $x$ as long as they were all greater than $1/2$. In each case we would get an infinite series summing to a constant. This demonstrates the power of a general equation to generate an infinite number of specific identities.

Our next example equates two infinite series, both involving the unknown $x$. Hence, this equation can also be used to generate additional equations. First we show it with both infinite series in compressed form, and then with each infinite series expanded.

$$\sum_{k=1}^{\infty} \left(1 + \frac{1}{2} + \frac{1}{3} + \ldots + \frac{1}{k}\right) \frac{x^k}{k!} = e^x \sum_{k=1}^{\infty} \frac{(-1)^{k-1} x^k}{k! k}$$

On the left of this equation, each term we are adding is the sum of a finite number of fractions multiplied by $x^k/k!$; there are an infinite number of such terms to add. When we expand both sides we get:

$$x + \frac{3x^2}{4} + \frac{11x^3}{36} + \frac{25x^4}{288} + \ldots = e^x \left[x - \frac{x^2}{4} + \frac{x^3}{18} - \frac{x^4}{96} + \ldots \right]$$

We see less symmetry within the expanded form of this equation, yet it is a marvel for no other reason than Ramanujan has equated two infinite series plus the constant $e$.

Now we pass on to an infinite series related to both a natural logarithm, but also to the Euler-Mascheroni constant which we introduced earlier. This constant, designated as $\gamma$ (Greek small gamma), is approximately equal to $0.5772157\ldots$. Remember that $\gamma$ was first discovered by Euler and relates the harmonic series ($\Sigma\, 1/n$) to the natural logarithms, or:

$$\gamma = \lim_{n \to \infty} \left( \sum_{j=1}^{n} \frac{1}{j} - \ln n \right)$$

Since both the harmonic series and natural logarithms are so important to mathematics, this constant which relates them is also important. Ramanujan's equation involving $\gamma$ is not an equality, but shows that an infinite series approximates the value of a logarithm plus the Euler-Mascheroni constant. In general, we prefer relationships in mathematics that are equalities, i.e., where the left and right side of the equation are exactly equal to each other. However, sometimes this is not possible, and we must be satisfied with an approximation. Such approximation relationships are useful when the approximation is close.

$$\sum_{k=1}^{\infty} \frac{(-1)^{k-1}x^k}{k!k} = x - \frac{x^2}{4} + \frac{x^3}{18} - \frac{x^4}{96} + \ldots \approx \ln x + \gamma$$

Again we notice the symmetry of alternating signs to individual terms and progressive powers of $x$.

## FRACTIONS AND RADICALS

We now shift gears and look at another area where Ramanujan excelled: infinite fractions. Our first example equates a simple infinite fraction to a constant.

$$\frac{4}{3} = \cfrac{3}{1 + \cfrac{4}{2 + \cfrac{5}{3 + \cfrac{6}{4 + \ldots}}}}$$

The symmetry in the above continued fraction is so elegant that the progressive terms are easily determined. The numerators, beginning with 3, increase through the number sequence, while the denominators, beginning with 1, do the same. Another example of such a continued fraction is:

$$\frac{5}{3} = \cfrac{4}{1 + \cfrac{6}{3 + \cfrac{8}{5 + \cfrac{10}{7 + \ldots}}}}$$

Here, the numerators are the even numbers beginning with 4 while the denominators are the odd numbers, beginning with 1.

In our next continued fraction we find the constant $e$.

$$\frac{1}{e-1} = \cfrac{1}{1 + \cfrac{2}{2 + \cfrac{3}{3 + \cfrac{4}{4 + \dots}}}}$$

This is an especially elegant continued fraction. The value of both sides of the equal sign is approximately 0.5819. Next we progress to an infinite continued fraction involving both the constant $e$ and the unknown, $x$.

$$\frac{e^x - 1}{e^x + 1} = \cfrac{x}{2 + \cfrac{x^2}{6 + \cfrac{x^2}{10 + \cfrac{x^2}{14 + \dots}}}}$$

This equation has exquisite symmetry with the numerators all $x^2$, except the first, and all the denominators differing by 4. However, what is nice about the equation is that it is one of the general equations we can use to derive additional equations by substituting in for $x$. We could, of course, substitute 1 or 2, but the results would not be any more spectacular than the original. However, if we substitute $\pi$ in for $x$ we get the beautiful equation:

$$\frac{e^\pi - 1}{e^\pi + 1} = \cfrac{\pi}{2 + \cfrac{\pi^2}{6 + \cfrac{\pi^2}{10 + \cfrac{\pi^2}{14 + \dots}}}}$$

Our next example is also a general form involving the unknown, $n$, which can assume the value of any positive integer.

$$n = \cfrac{1}{1 - n + \cfrac{2}{2 - n + \cfrac{3}{3 - n + \dots \cfrac{n}{0 + \cfrac{n+1}{1 + \cfrac{n+2}{2 + \dots}}}}}}$$

This is a strange continued fraction, indeed. The first ellipsis that occurs after the $3 - n$ term means that we continue to subtract $n$ from larger and larger integers until the sum is zero. From then on, we simply use the positive integer sequence. In other words, if we let $n$ be some positive integer, then in the $n$th denominator in the continued fraction we will have $n - n = 0$, which is how we get the zero in the above equation. We might object that it is not possible to have a denominator that is zero, but remember that we are adding something to this zero, so that the denominator has a value other than zero. For example, let's substitute 3 in for $n$.

$$3 = \cfrac{1}{-2 + \cfrac{2}{-1 + \cfrac{3}{0 + \cfrac{4}{1 + \ldots}}}}$$

Next we have an equation involving the unknown $x$ where $x$ can be any number except a negative integer.

$$1 = \cfrac{x + 1}{x + \cfrac{x + 2}{x + 1 + \cfrac{x + 3}{x + 2 + \cfrac{x + 4}{x + 3 + \ldots}}}}$$

From this equation we can generate any number of additional infinite continued fractions all equal to 1.

Our next infinite continued fraction is related to an infinite series.

$$2 \cdot \sum_{k=1}^{\infty} \frac{(-1)^{k+1}}{x + 2k - 1} = \cfrac{1}{x + \cfrac{1^2}{x + \cfrac{2^2}{x + \cfrac{3^2}{x + \cfrac{4^2}{x + \ldots}}}}}$$

When we expand the infinite series we get the beautifully symmetric equation:

$$2\left(\frac{1}{x+1} - \frac{1}{x+3} + \frac{1}{x+5} - \frac{1}{x+7} + \cdots\right) = \cfrac{1}{x + \cfrac{1^2}{x + \cfrac{2^2}{x + \cfrac{3^2}{x + \cfrac{4^2}{x + \cdots}}}}}$$

We can substitute any number for $x$ as long as it is greater than zero. Even substituting a simple 1 for $x$ yields the following pleasant identity.

$$2\left(\frac{1}{2} - \frac{1}{4} + \frac{1}{6} - \frac{1}{8} + \cdots\right) = \cfrac{1}{1 + \cfrac{1}{1 + \cfrac{4}{1 + \cfrac{9}{1 + \cfrac{16}{1 + \cdots}}}}}$$

This next equation contains not only an infinite series and a continuing fraction, but both the constants $\pi$ and $e$.

$$1 + \frac{x}{3} + \frac{x^2}{3 \cdot 5} + \frac{x^3}{3 \cdot 5 \cdot 7} + \frac{x^4}{3 \cdot 5 \cdot 7 \cdot 9} = \sqrt{\frac{\pi}{2x}} \cdot e^{\frac{x}{2}} - \cfrac{1}{x + \cfrac{1}{1 + \cfrac{2}{x + \cfrac{3}{1 + \cfrac{4}{x + \cdots}}}}}$$

Symmetry can be seen everywhere in this equation. We have the numerators of the infinite series as progressive powers of $x$, while the denominators are products of successive odd numbers. In the continuing fraction the numerators are successive integers after the initial two 1s. The denominators alternate between $x$ and 1.

We're going to change forms again and look at continuing radicals. One of Ramanujan's general forms for continuing radicals was the complex looking:

$$x - a_1 = \sqrt{x^2 + a_1(a_1 - 2a_2) - 2a_1\sqrt{x^2 + a_2(a_2 - 2a_3) - 2a_2\sqrt{x^2 + \cdots}}}$$

The pattern within the radicals is a little hard to catch. Notice that each radical contains three terms, one involving $x$ and two involving an $a$ with a subscript. The first term is always $x^2$, the second term is $a_1(a_1 - 2a_2)$ except the subscripts increase by one for each new radical. The third term is $2a_1$ and the subscript increases by one for each new radical. What is truly unusual about this continuing radical is that the identity is true for all $x$ and for all sets of $a_1$, $a_2$, $a_3$, etc. By making specific substitutions for $x$ and the $a$s we can get some beautiful continued radicals. For example, we'll let $x = 1$ and $a_1 = 1/2$, $a_2 = 1/4$, $a_3 = 1/8$, etc. Making the substitutions and solving for the continued radical we get:

$$\frac{1}{2} = \sqrt{1 - \sqrt{1 - \frac{1}{2}\sqrt{1 - \frac{1}{4}\sqrt{1 - \frac{1}{8}\sqrt{1 - \ldots}}}}}$$

We can generate another equation by substituting $x = 3$, $a_1 = -1$, $a_2 = 1$, $a_3 = -1$, $a_4 = 1$, etc. This yields:

$$4 = \sqrt{12 + 2\sqrt{12 - 2\sqrt{12 + 2\sqrt{12 - \ldots}}}}$$

When making substitutions for $x$ and $a_1$, $a_2$, etc., we must make sure that $x$ is larger than $a_1$ or the resulting continued radical will involve complex numbers.

Remember that we mentioned Ramanujan's equation appearing in the *Journal of the Indian Mathematical Society*?

$$3 = \sqrt{1 + 2\sqrt{1 + 3\sqrt{1 + 4\sqrt{1 + \ldots}}}}$$

This equation was generated from the following general form which Ramanujan had discovered.

$$x + n + a = \sqrt{ax + (n + a)^2 + x\sqrt{a(x + n) + (n + a)^2 + (x + n)\sqrt{\ldots}}}$$

The pattern within this equation is also somewhat difficult to see. We have three basic terms under each radical. The first term has the following progression: $ax$, $a(x + n)$, $a(x + 2n)$, $a(x + 3n)$, $a(x + 4n)$, etc. The second term, $(n + a)^2$, does not change while the third term progresses as: $x$, $x + n$, $x + 2n$, $x + 3n$, etc. For example, we can let $x = 1$, $n = 2$, and $a = 3$. This yields:

$$6 = \sqrt{28 + \sqrt{34 + 3\sqrt{40 + 5\sqrt{46 + \ldots}}}}$$

We might wonder if any substitution of numbers into an infinite continued radical will result in a finite number or limit, or will some substitutions give us an expression that is unbounded? In other words, will infinite continued radicals always converge to a limit? It was discovered by T. Vijayaraghavan that the infinite radical,

$$\sqrt{a_1 + \sqrt{a_2 + \sqrt{a_3 + \sqrt{a_4 + \ldots}}}}$$

where $a_n \geq 0$, will converge to a limit if and only if the limit of $(\ln a_n)/2^n$ exists. Hence, if $(\ln a_n)/2^n$ does exist for a specific sequence, when the sequence is substituted into an infinite radical, it will also have a limit or specific value.

We are not always restricted to square roots as radicals. For example, we have the following two cube root radicals.

$$-2 = \sqrt[3]{-6 + \sqrt[3]{-6 + \sqrt[3]{-6 + \sqrt[3]{-6 + \ldots}}}}$$

$$1 = \sqrt[3]{3 + \sqrt[3]{-6 + \sqrt[3]{-6 + \sqrt[3]{-6 + \ldots}}}}$$

While we cannot have a real negative number as a square root (unless we are willing to deal with complex numbers), we can have negative cube roots. This is easily seen when we consider the cube root of $-8$ which is $-2$, or $(-2)(-2)(-2) = -8$.

The last three examples of Ramanujan's equations we will consider are very special, for they deal with prime numbers. We now know that such equations are difficult to find and, once found, frequently reveal deep secrets of the number sequence. For our next equation we will need to define a new symbol. When we wish to add a sequence of terms together we use a large Greek sigma ($\Sigma$). When we want to show a sequence of terms all multiplied together we use a large Greek pi ($\Pi$). Hence to multiply all the positive integers together from 1 to $n$ as one product, we could show it as:

$$\prod_{k=1}^{n} k = 1 \cdot 2 \cdot 3 \cdot \ldots \cdot n$$

This looks just like the factorial of $n$, but we can actually display much more complex functions using this notation because we can replace $k$ with a more complex expression. This is just what we have in the next Ramanujan equation.

$$\prod_{p}^{\infty} \left( \frac{p^2 + 1}{p^2 - 1} \right) = \frac{5}{2}$$

In this equation $p$ stands for the sequence of all prime numbers. Hence, an expanded version of this equation would be:

$$\left( \frac{2^2 + 1}{2^2 - 1} \right) \cdot \left( \frac{3^2 + 1}{3^2 - 1} \right) \cdot \left( \frac{5^2 + 1}{5^2 - 1} \right) \cdots = \left( \frac{5}{3} \right) \left( \frac{10}{8} \right) \left( \frac{26}{24} \right) \cdots = \frac{5}{2}$$

A second equation dealing with the sequence of prime numbers is:

$$\prod_{p}^{\infty} \left( 1 + \frac{1}{p^4} \right) = \left( 1 + \frac{1}{2^4} \right) \left( 1 + \frac{1}{3^4} \right) \left( 1 + \frac{1}{5^4} \right) \left( 1 + \frac{1}{7^4} \right) \cdots = \frac{105}{\pi^4}$$

This remarkable equation relates the prime number sequence on the left to $\pi$ on the right. How can prime numbers be related to the ratio of a circle's circumference to its diameter? This is yet another wonderful example of the interrelatedness of mathematics.

In our final example, we have another approximation rather than an equality. If we let $P_k$ be the $k$th prime, then as $x$ approaches zero, we have:

$$\sum_{k=1}^{\infty} \frac{P_k}{e^{kx}} = \frac{2}{e^x} + \frac{3}{e^{2x}} + \frac{5}{e^{3x}} + \frac{7}{e^{4x}} + \cdots \approx -\frac{\ln x}{x^2}$$

In this example, the sequence of prime numbers is related to not only the constant $e$ but also to $\ln x$. For Ramanujan to have discovered such beautiful relationships between $e$, logarithms, and primes shows the great depth of his genius.

## PARTITIONS

During his stay at Cambridge Ramanujan contributed to a field of number theory called partitions. What begins as a simple question quickly turns into a most difficult problem. A partition of a number is just an expression of the number as a sum of numbers less than or equal to the given number. For example, $4 = 1 + 1 + 1 + 1$, and $4 = 3 + 1$ are two different parts of 4 while $5 = 3 + 2$ and $5 = 1 + 2 + 2$ are two different parts of 5. The basic question is: How many partitions of a given number are there? Let's consider all the ways of combining natural numbers to get 4.

$$4 = 1 + 1 + 1 + 1$$

$$4 = 2 + 1 + 1$$

$$4 = 3 + 1$$

$$4 = 2 + 2$$

$$4 = 4$$

Hence, we see that if we don't consider the order of the terms, then we can write 4 as sums of numbers that are equal to or smaller than 4 in five different ways and that is all. There are five parts of 4, and 5 is called the partition number of 4. Generally we designate the partition number as $p(n)$, or $p(4) = 5$. You might hope that some easy formula will compute the partition numbers for the positive integers. However, this is not the case. The partition numbers increase quickly as $n$ grows larger and larger. For example, we have listed the partition numbers, $p(n)$, for the first 50 integers in Table 14. By the time we get to 100, $p(100) = 190,569,292$. The obvious question is: For any $n$, can we quickly compute $p(n)$?

Euler was the first to make any progress on this problem.[4] He proposed finding a function, $f(x)$, that generated an infinite series consisting of successive powers of $x$. The coefficients of the different powers of $x$ would be our $p(n)$. On first encounter, generating functions are a bit confusing. Just think of a generating function as a kind of mathematical engine that keeps spitting out successively greater powers of $x$. The coefficients connected to the various powers of $x$ are the numbers we are interested in. We can represent this idea symbolically as:

$$f(x) = 1 + \sum p(n)x^n = 1 + p(1)x^1 + p(2)x^2 + p(3)x^3 + \ldots$$

However, Euler was not able to find the proper function $f(x)$ that would generate the series desired. Hence, the problem lay dormant

**Table 14.** Partition Numbers for 1 through 50

| Number | Partition Number | Number | Partition Number |
|--------|------------------|--------|------------------|
| 1 | 1 | 26 | 2,436 |
| 2 | 2 | 27 | 3,010 |
| 3 | 3 | 28 | 3,718 |
| 4 | 5 | 29 | 4,565 |
| 5 | 7 | 30 | 5,604 |
| 6 | 11 | 31 | 6,842 |
| 7 | 15 | 32 | 8,349 |
| 8 | 22 | 33 | 10,143 |
| 9 | 30 | 34 | 12,310 |
| 10 | 42 | 35 | 14,883 |
| 11 | 56 | 36 | 17,977 |
| 12 | 77 | 37 | 21,637 |
| 13 | 101 | 38 | 26,015 |
| 14 | 135 | 39 | 31,185 |
| 15 | 176 | 40 | 37,338 |
| 16 | 231 | 41 | 44,583 |
| 17 | 297 | 42 | 53,174 |
| 18 | 385 | 43 | 63,261 |
| 19 | 490 | 44 | 75,175 |
| 20 | 627 | 45 | 89,134 |
| 21 | 792 | 46 | 105,558 |
| 22 | 1,002 | 47 | 124,754 |
| 23 | 1,255 | 48 | 147,273 |
| 24 | 1,575 | 49 | 173,525 |
| 25 | 1,958 | 50 | 204,226 |

until Hardy and Ramanujan began work on it. Ramanujan, while still in India, had derived a generating function which came close to giving good approximations to the $p(n)$ coefficients in the series. Using this function as a starting place, Hardy and Ramanujan went to work to improve the accuracy of the approximations.

They began this work in 1916, before the development of modern computers, or even hand calculators. Therefore, to test the accuracy of their improved generating functions, Hardy and Ramanujan had to know the actual values for lots of $p(n)$s and these had to be calculated—a rather daunting task at that time. To help in this effort, Hardy recruited the mathematician Percy MacMahon, a 61-year-old former Royal artilleryman, who was professor at Woolwich College. MacMahon was known for his ability to complete difficult calculations. Would he, Hardy asked, just compute $p(n)$ for the first 200 integers? MacMahon did and found that $p(200)$ = 3,972,999,029,388 which is just under four trillion, almost our national debt!

If MacMahon wrote down all the different ways of writing integers to equal 200, he would end up with the gigantic number above. If he were to add together all these almost four trillion ways at one second for each addition, it would take him 125,982 years just to do that! Obviously, MacMahon needed some help to compute these first 200 $p(n)$s. Lucky for him, there is a regression formula for computing $p(n)$. If you have already computed all the partition numbers up to $p(n-1)$, then it is relatively easy to get $p(n)$. The regression formula is:[5]

$$p(n) = p(n-1) + p(n-2) - p(n-5)$$
$$- p(n-7) + p(n-12) + p(n-15) - \ldots$$

This series is extended as long as the numbers within the parentheses are positive or zero. We define $p(0) = 1$. For example, suppose we had already computed $p(n)$ for 1 through 15, and now wanted $p(16)$. Using the above formula we get:

$$p(16) = p(15) + p(14) - p(11) - p(9) + p(4) + p(1)$$

or

$$p(16) = 176 + 135 - 56 - 30 + 5 + 1 = 231$$

We don't have to include any more terms because the next term in the expansion of the formula would result in a negative number within the parentheses. The formula is a little tricky to decipher. Notice that the terms can be grouped in pairs, each pair having the same sign; the first pair is added, the next pair subtracted, and so on.

$$p(n) = p(n-1) + p(n-2)$$
$$- p(n-5) - p(n-7)$$
$$+ p(n-12) + p(n-15) \ldots$$

The differences within each pair increase by one for each successive pair; hence, the first pair differ by 1 (1 and 2), the second pair by 2 (5 and 7), the third pair by 3 (12 and 15), etc. The differences between the second element of a preceding pair, and the first element of a succeeding pair increase by 2 between each pair, starting with the first difference of 3. For example, the different between $p(n - 2)$ and $p(n - 5)$ is 3, while the difference between $p(n - 7)$ and $p(n - 12)$ is 5. Hence, the differences between pairs are 3, 5, 7, 9, 11, etc.

While this formula works fine for computing the first 200 values of $p(n)$, it suffers from several serious drawbacks. First, since it depends on all previous $p(j)$ where $j < n$, if we make a mistake at $n = 12$, then every $p(n)$ we compute after 12 will also be in error. Therefore, it is important we make no mistakes to carry forward, making all the rest of the answers invalid. Second, if we want to compute large partition values, the formula is difficult to use. For example, using it to compute $p(1000)$ is a daunting task. It would be preferable to have a generating function such as Euler suggested that would directly estimate $p(1000)$ without computing the 999 $p(n)$s that come before $p(1000)$.

This is what Hardy and Ramanujan were after. Finally in December 1916 Ramanujan was able to improve their function to get very accurate estimates for $p(n)$. Their approximating formula[6] is a series whose first term is:

$$p(n) \approx \frac{e^{\pi\sqrt{\frac{2n}{3}}}}{4n\sqrt{3}} + \dots$$

Using just this first term we can get fair approximations of $p(n)$ because succeeding terms are relatively small and get progressively smaller. For example, using just the first term we get $p(200) \approx 4.1 \times 10^{12}$, an error from the true value of $p(200)$ of approximately 3.3%. Including additional terms in Hardy and Ramanujan's approximation formula produces results accurate to within one integer of the true value of $p(n)$.

Ramanujan was not satisfied with just getting an extremely accurate estimate of $p(n)$, but went on to discover some congruence relationships between $n$ and $p(n)$—relationships which had never been suspected before. What he discovered was that if we take any $n$ which has a remainder of 4 when divided by 5 (or $n \equiv 4 \bmod 5$), then $p(n)$ is evenly divisible by 5, or $p(n) \equiv 0 \bmod 5$. This whole thing can be written as $p(5n + 4) = 0 \bmod 5$ for all integers, $n$, which are equal to zero or greater. He also discovered relationships for 7 and 11:

$$p(7n + 5) \equiv 0 \bmod 7 \quad \text{and} \quad p(11n + 6) \equiv 0 \bmod 11$$

One may wonder what in the world any of this partitions stuff could have to do with real-world problems. Hardy would probably turn over in his grave if he knew that today the theory of partitions is used in solving practical problems in such diverse fields as communication lines and plastics. Physicists are using partition theory to solve problems from statistical mechanics to string theory—a theory concerning the unification of matter and energy in the universe.[7]

## CHAPTER TWELVE

# GOLDBACH'S CONJECTURE

*The principal difficulty in the mathematics is the*

*length of inferences and compass of thought requisite*

*to the forming of any conclusion.*

David Hume (1711–1776)

*An Enquiry Concerning Human Understanding*[1]

## ADDING IT ALL UP

Sometimes the most innocent question inspires the greatest effort in mathematics. Christian Goldbach (1690–1764) asked just such a question in 1742. Goldbach was a German mathematician who became professor of mathematics in 1725 in St. Petersburg, Russia. Three years later he traveled to Moscow to tutor Tsar Peter II. Goldbach traveled about Europe during his career and met a number of talented mathematicians, among them Leonhard Euler. Along with Goldbach, Euler became a member of the Russian Academy of Sciences. Both men had a love for infinite series and prime numbers. In a letter to Euler on June 7, 1742, Goldbach speculated that every even number is the sum of two primes, and every odd number larger than 2 is the sum of three primes.[2]

At that time it was not clear whether 1 should be considered a prime, and in his letter, Goldbach was assuming it was. Since we now exclude 1 as a prime, the modern statements of Goldbach's Conjectures are: Every even number 4 and greater can be expressed as the sum of two primes, and every odd number 7 and greater can be expressed as the sum of three primes. The first part of this claim is called the Binary Goldbach Conjecture, and the second part is the Ternary Goldbach Conjecture.

After all these years, the Binary Goldbach Conjecture is still not proven, even though virtually all mathematicians believe it is true, and countless hours have been spent on its solution. For such a simple question, Goldbach certainly stumped thousands of mathematicians for 250 years!

Goldbach's Ternary Conjecture has been proven—almost! In 1923 G. H. Hardy and John Littlewood proved that there exists a number, $n$ (which they didn't know), such that every odd number larger then $n$ can be written as the sum of three prime numbers. However, to prove this they had to assume as true a general form of the Riemann hypothesis, another famous conjecture that has not been proven true to this day. All that Hardy and Littlewood proved was that *if* the Riemann hypothesis is true, then beyond some number $n$ all odd numbers were the sum of three primes. This meant that if the Riemann hypothesis could be proven, then the Ternary Conjecture could only fail to be true for a finite number of odd numbers, e.g., some set of odd numbers less than $n$.

This wasn't really very close, yet Hardy and Littlewood were on the right track. And they did make progress after 180 years of no progress at all. In fact, it was Hardy and Littlewood who defined an entirely new approach to studying problems involving sums of numbers, known as additive number theory. Then in 1937, I.M. Vinogradov proved the same thing as Hardy and Littlewood, except he didn't have to assume the Riemann hypothesis was true. Hence, we now know that there exists some number such that all odd numbers that are larger can be written as the sum of three primes. This reduces the problem to finding this number $n$, and then testing all odd primes up to $n$ to verify that they, too, can be written as the sum of three primes. (Hopefully, $n$ is not too big!) This will settle once and for all that the Ternary Conjecture is true. But there is still that little way to go before we can claim total victory. After all, we may test all odd numbers up to $n$, only to discover one or more cannot be written as the sum of three primes.

How big is $n$? One of the first estimates of its size was approximately:[3]

$$3^{3^{15}} = 3^{14,348,907} \approx 10^{6,846,168}$$

To say the least, this is a rather large number, approximately ten followed by 6,846,000 zeros. To test all odd numbers up to this limit would take more time and computer power than we have. Recent work has improved the estimate of $n$. In 1989 J.R. Chen and T. Wang computed $n$ to be approximately:[4]

$$e^{e^{11.503}} \approx 10^{43,000}$$

This new value for $n$ is much smaller than the previous one, and suggests that some day soon we will be able to test all odd numbers up to this limit to see if they can be written as the sum of three primes.

## GOLDBACH'S BINARY CONJECTURE

The whole idea seems so simple: Can every even number, 4 or greater, be written as the sum of two primes? Let's look at the first even numbers:

$4 = 2 + 2$
$6 = 3 + 3$
$8 = 3 + 5$
$10 = 3 + 7$ and $5 + 5$
$12 = 5 + 7$
$14 = 3 + 11$ and $7 + 7$
$16 = 3 + 13$ and $5 + 11$
$18 = 5 + 13$ and $7 + 11$
$20 = 3 + 17$ and $7 + 13$

It would appear on the surface that Goldbach's Conjecture is true, and that as even numbers get larger, the number of ways of writing an even number as two primes increases. In fact, the very next number, 22, can be written as the sum of two primes in three different ways: $3 + 19, 5 + 17$, and $11 + 11$. The numbers 4, 6, 8, and

12 are the only numbers known that can be written as the sum of two primes in only one way.

The various attacks on this problem have been interesting. Suppose that the number of even integers less than $n$ that cannot be written as the sum of two primes is $F(n)$. Of course we would like to prove that $F(n) = 0$ for all $n$; that is, there are no even integers that cannot be so written. In the late 1930s, T. Estermann, J.G. van der Corput, and Chudakov proved that the ratio of $F(n)/n$ approaches zero as $n$ goes to infinity. This is far from proving that $F(n)$ is zero for all $n$, but it did demonstrate that if there did exist even numbers that failed the Binary Conjecture, their number was small compared to $n$. Mathematicians would say that the density of $F(n)$ is very low. An example will illustrate the idea of density. Suppose we consider the density of all even numbers among the integers. Up to and including the number $2n$ we know there are $n$ even numbers. Hence, the ratio of even numbers to all numbers up to $2n$ is just $n/2n = 1/2$ or 50% Hence, no matter how big a number we choose, we know the density of even numbers will always be 50%.

Another way to attack the Binary Conjecture is to start by proving all even numbers can be written as the sum of two numbers, these two numbers being composed of no more than $n$ and $m$ primes. Hence, for any even number $N$, we have:

$$N = P_a \cdot P_b \cdot P_c \cdot \ldots \cdot P_n + P_A \cdot P_B \cdot P_C \cdot \ldots \cdot P_m$$

For example, let $N = 56$ and $n = 3$ while $m = 2$. We can write 56 as the sum of the product of three primes ($n$) plus the product of two primes ($m$) or:

$$56 = 2 \cdot 3 \cdot 5 + 2 \cdot 13 = 30 + 26$$

Now, what we would like to prove is that for all even $N$, $n$ and $m$ can both be 1; that is, that both numbers we add contain only one prime. When we reach this point, the two terms combining to make the sum will each be a single prime and the conjecture will be proven. In 1919 V. Brun proved that every sufficiently large even number could be written as the sum of two products, each product consisting of no more than nine primes. We show this as

$$N = P_9 + P_9$$

Here the subscripts show the maximum number of primes contained in the number $P$. No advance was made for a number of years. Then, in 1950 the team consisting of Rademacher, Estermann, Ricci, Buchstab, and Selberb proved that all sufficiently large even primes could be written as

$$N = P_2 + P_3$$

Getting closer! Finally, in 1966 J.R. Chen[5] proved that every sufficiently large even number could be written as

$$N = P_1 + P_2$$

Really close! This says that every sufficiently large even number can be written as the sum of a prime and a near prime, i.e., a number composed of only two primes. Now all we have to do is reduce the equation from $N = P_1 + P_2$ to $N = P_1 + P_1$, then find out how large "sufficiently large" is, and finally test all even numbers up to that number.

Of course, there have been attempts to test even numbers to find one that cannot be written as the sum of two primes. This would disprove the Binary Conjecture. By 1993 the even numbers up to $4 \times 10^{11}$ or 400,000,000,000 had been tested with no such counterexample being found.[6]

## GOLDBACH'S COMET

There is yet another way to look at Goldbach's Binary Conjecture thanks to the work of Henry Fliegel, Aerospace Corporation, and Douglas Robertson, National Geodesic Survey.[7] Fliegel and Robertson have expanded the idea behind the Goldbach Conjecture in an intriguing way. Let $n$ be an integer, and let $C(n)$ be the number of ways $n$ can be written as the sum of two primes. Hence, we have $C(4) = 1$ because 4 can be written as the sum of two primes in only one way: $2 + 2$. On the other hand, $C(34) = 4$ because 34 can be written as: $3 + 31$, $5 + 29$, $11 + 23$, and $17 + 17$. The numbers, $C(n)$, are called Goldbach numbers. Now we can investigate how Goldbach numbers change as $n$ increases through the integers. Of

**Table 15.** Goldbach Numbers for Even Numbers, 2 through 200

| n | C(n) | n | C(n) | n | C(n) | n | C(n) | n | C(n) |
|---|------|---|------|---|------|---|------|---|------|
| 2 | 0 | 42 | 4 | 82 | 5 | 122 | 4 | 162 | 10 |
| 4 | 1 | 44 | 3 | 84 | 8 | 124 | 5 | 164 | 5 |
| 6 | 1 | 46 | 4 | 86 | 5 | 126 | 10 | 166 | 6 |
| 8 | 1 | 48 | 5 | 88 | 4 | 128 | 3 | 168 | 13 |
| 10 | 2 | 50 | 4 | 90 | 9 | 130 | 7 | 170 | 9 |
| 12 | 1 | 52 | 3 | 92 | 4 | 132 | 9 | 172 | 6 |
| 14 | 2 | 54 | 5 | 94 | 5 | 134 | 6 | 174 | 11 |
| 16 | 2 | 56 | 3 | 96 | 7 | 136 | 5 | 176 | 7 |
| 18 | 2 | 58 | 4 | 98 | 3 | 138 | 8 | 178 | 7 |
| 20 | 2 | 60 | 6 | 100 | 6 | 140 | 7 | 180 | 14 |
| 22 | 3 | 62 | 3 | 102 | 8 | 142 | 8 | 182 | 6 |
| 24 | 3 | 64 | 5 | 104 | 5 | 144 | 11 | 184 | 8 |
| 26 | 3 | 66 | 6 | 106 | 6 | 146 | 6 | 186 | 13 |
| 28 | 2 | 68 | 2 | 108 | 8 | 148 | 5 | 188 | 5 |
| 30 | 3 | 70 | 5 | 110 | 6 | 150 | 12 | 190 | 8 |
| 32 | 2 | 72 | 6 | 112 | 7 | 152 | 4 | 192 | 11 |
| 34 | 4 | 74 | 5 | 114 | 10 | 154 | 8 | 194 | 7 |
| 36 | 4 | 76 | 5 | 116 | 6 | 156 | 11 | 196 | 9 |
| 38 | 2 | 78 | 7 | 118 | 6 | 158 | 5 | 198 | 13 |
| 40 | 3 | 80 | 4 | 120 | 12 | 160 | 8 | 200 | 8 |

course, if we ever find an $n$ such that $C(n) = 0$, then we will have disproved the Binary Conjecture. Table 15 lists the Goldbach numbers for the first 200 even integers. By examination of Table 15 we realize that Goldbach numbers are increasing with increasing $n$, yet within any short interval they seem to jump around in an apparently random manner.

Fliegel and Robertson computed $C(n)$ for hundreds of thousands of numbers and then plotted their values. The results contained a completely unexpected surprise. Figure 42 is the plot of 5000 Goldbach numbers through 10,000. The horizontal scale is the integer $n$, and the vertical scale represents the values of $C(n)$. Notice that the values of the Goldbach numbers do not spread out evenly over their range, but tend to cluster into definite bands. Certainly, something strange is going on here. Because of the overall shape of this graph, Fliegel and Robertson named it a Goldbach Comet.

C(n)

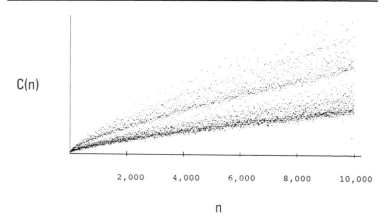

2,000        4,000        6,000        8,000        10,000

n

FIGURE 42. Goldbach's Comet. C(n) is the number of ways of representing n, an even number, as the sum of two prime numbers. As n increases, C(n) breaks into bands.

In addition to the bands, there are several remarkable features of the Goldbach Comet. The lowest set of points, that is, the smallest Goldbach numbers, form almost a sharp edge along the lower part of the comet. This suggests that not only is the Binary Conjecture true for all even numbers, but there is a lower bound for Goldbach numbers that steadily increases as n increases. Checking specific values for Goldbach numbers reinforces this idea. Only four numbers are known to have a C(n) value of 1: 4, 6, 8, and 12. No number larger than 632 is known to have a Goldbach number smaller than 10, and no number larger than 1448 is known to have a Goldbach number smaller than 20. This is evidence for the truth of the conjecture. Had we seen a smattering of values from one up in all sections of the graph, we might not be so confident. On the other hand, we do see a smattering of large values at the top of the graph, which suggests that there may be no limit to the number of ways large enough numbers can be expressed as the sums of primes.

While this graphical evidence is suggestive, it does not prove the conjecture. But what causes the bands? These bands appear at heights whose ratios are simple fractions from the height of the lowest band. By using the theory of congruences, Fliegel and

**Table 16.** Six-Column Array of Numbers

| Column 1 | Column 2 | Column 3 | Column 4 | Column 5 | Column 6 |
|----------|----------|----------|----------|----------|----------|
| 1 | 2 | 3 | 4 | 5 | 6 |
| 7 | 8 | 9 | 10 | 11 | 12 |
| 13 | 14 | 15 | 16 | 17 | 18 |
| 19 | 20 | 21 | 22 | 23 | 24 |
| 25 | 26 | 27 | 28 | 29 | 30 |
| 31 | 32 | 33 | 34 | 35 | 36 |
| 37 | 38 | 39 | 40 | 41 | 42 |
| 43 | 44 | 45 | 46 | 47 | 48 |
| 49 | 50 | 51 | 52 | 53 | 54 |
| 55 | 56 | 57 | 58 | 59 | 60 |
| 61 | 62 | 63 | 64 | 65 | 66 |
| 67 | 68 | 69 | 70 | 71 | 72 |
| 73 | 74 | 75 | 76 | 77 | 78 |
| 79 | 80 | 81 | 82 | 83 | 84 |
| 85 | 86 | 87 | 88 | 89 | 90 |
| 91 | 92 | 93 | 94 | 95 | 96 |
| 97 | 98 | 99 | 100 | 101 | 102 |
| 103 | 104 | 105 | 106 | 107 | 108 |
| 109 | 110 | 111 | 112 | 113 | 114 |
| 115 | 116 | 117 | 118 | 119 | 120 |

Robertson were able to solve the mystery. Remember that congruence theory is simply a way to describe what happens when we write numbers in fixed columns. For example, Table 16 shows a six-column array of integers from 1 to 120. Notice that if we divided any number in column 5 by the number 6 (the array size), we get a remainder of 5. On the other hand, if we divide a number from column 1 by 6 we get a remainder of 1. Hence, the column number tells us the remainder we get by dividing a number by the size of the array. The only exception, of course, is column 6 where we get a remainder of zero if we divide by 6.

We symbolically show a congruence relationship in the following manner: $25 \bmod 6 \equiv 1$, where the three bars is a congruence sign, and we read the relationship as "25 is congruent to 1 modulo 6." This means two things. First, the number 25 is in the first (number 1) column, and that if we divide 25 by 6 the remainder is 1. We have already pointed out that interesting things happen to numbers when they are arranged in arrays. For example, in our six-column array, all prime numbers fall into columns 1 and 5, and no others.

How can we use congruences to understand the bands in Goldbach's Comet? A law of congruences states the following:

$$\text{if } A = B + C \text{ then } A \bmod m = B \bmod m + C \bmod m$$

A little thought will illustrate that this is so, for $A \bmod m$ is the remainder we get when we divide $A$ by $m$. Now let's say that $A$ is 43 and $m$ is 5. This means that $43 \bmod 5 = 3$, because the remainder of dividing 43 by 5 is 3. Now we will write 43 as the sum of two numbers: $43 = 31 + 12$. If we divide both 31 and 12 by 5, the sum of those remainders must equal 3, or $31 \bmod 5 = 1$, $12 \bmod 5 = 2$, and $1 + 2 = 3$.

A Goldbach number, $C(n)$, represents the number of ways we can write an even number as the sum of two primes. If $N = P_a + P_b$, then $N \bmod m = P_a \bmod m + P_b \bmod m$. This says that if we divide $N$ by $m$ to get a remainder, then this remainder will equal the sum of the remainders we get when we divide the two primes by the same $m$. Now let's see what happens when we compute the Goldbach numbers for the even numbers found in Table 16. These Goldbach numbers are found in Table 17, where the entries represent the number of ways the even numbers from Table 16 can be written as the sum of two primes. (Notice we have entered a zero for the number 2 since it cannot be written as the sum of two primes.) We only show the Goldbach numbers for columns 2, 4, and 6 in Table 17 because we are only calculating the Goldbach numbers for the even numbers from Table 16. A simple inspection of Table 17 shows that the Goldbach numbers in column 6 are generally larger than those in columns 2 and 4. We can confirm this by adding the three columns.

The sum of the Goldbach numbers from column 2 is 65, the sum from column 4 is 82, and the sum from column 6 is 115. Why this difference, and is it just an accident—a random occurrence? Table 17 suggests that numbers in column 6 of Table 16 can be written as the sum of two primes in more ways, on average, than the numbers from the other two columns. Remember that each number in the

**Table 17.** Goldbach Numbers in Six-Column Array

| Column 1 | Column 2 | Column 3 | Column 4 | Column 5 | Column 6 |
|---|---|---|---|---|---|
|  | 0 |  | 1 |  | 1 |
|  | 1 |  | 2 |  | 1 |
|  | 2 |  | 2 |  | 2 |
|  | 2 |  | 3 |  | 3 |
|  | 3 |  | 2 |  | 3 |
|  | 2 |  | 4 |  | 4 |
|  | 2 |  | 3 |  | 4 |
|  | 3 |  | 4 |  | 5 |
|  | 4 |  | 3 |  | 5 |
|  | 3 |  | 4 |  | 6 |
|  | 3 |  | 5 |  | 6 |
|  | 2 |  | 5 |  | 6 |
|  | 5 |  | 5 |  | 7 |
|  | 4 |  | 5 |  | 8 |
|  | 5 |  | 4 |  | 9 |
|  | 4 |  | 5 |  | 7 |
|  | 3 |  | 6 |  | 8 |
|  | 5 |  | 6 |  | 8 |
|  | 6 |  | 7 |  | 10 |
|  | 6 |  | 6 |  | 12 |
| Column Sums | 65 |  | 82 |  | 115 |

even columns of Table 16 must be written as the sum of two primes, and all primes must come from columns 1 and 5 of Table 16, for these are the only two columns containing primes. Every number in column 2 has a remainder of 2 when divided by 6, or $N_2 \bmod 6 = 2$. Therefore, if we write the number from column 2, Table 16, as the sum of two primes we have: $N_2 = P_a + P_b$. We now ask where the two primes $P_a$ and $P_b$ can come from. If we divide $P_a$ and $P_b$ by 6, the sum of the remainders must equal 2 since $N_2 \bmod 6 = 2$. The only column we can get these two primes from is column number 1. We can verify this by checking Table 16. If each prime is from column 1 then the two remainders after dividing by 6 will be 1, and therefore, the sum of the remainders will be 2. On the other hand, if we get one prime from column 1 and one prime from column 5, then the sum of the remainders will be $1 + 5 = 6$ (or 0), not 2. If we get both from column 5, the sum of the remainders will be 10 which is equivalent to column 4 (after dividing by 6). Hence, the primes can't both come from column 5. The only possibility is that both primes come from column 1 so that the sum of their remainders is equal to 2. Hence, in choosing two primes to add together to yield a number in column 2, we can use only those primes from column 1.

Now let's consider those numbers in column 6. For every number in column 6, $N_6 \bmod 6 = 0$, or the remainder when dividing by 6 is always zero. Which of the two columns, 1 and 5, can contribute primes to add to $N_6$? In fact, a prime from column 1 will give us a remainder of 1 and a prime from column 5 will give us a remainder of 5. The sum of these two remainders is 6, which is equivalent to zero, modulo 6. Hence, in adding primes to sum to $N_6$, we can try one prime from column 1 and one from column 5. Therefore, there are more possible combinations that might add to $N_6$ when we try combinations of primes from both columns rather than being restricted to just one column. This is why the number of ways of writing $N_6$ as the sum of two primes in column 6 is greater than those numbers from column 2.

A concrete example helps. The number 36 is in column 6 of Table 16. There are four ways primes can be added together to get 36, namely: $5 + 31, 7 + 29, 13 + 23$, and $17 + 19$. Notice in each case, one prime comes from column 1 and the other from column 5. Now the number 38 can be written as the sum of two primes in just two ways: $7 + 31$ and $19 + 19$. In each case both primes had to come from column 1.

The same analysis applies to column 4. When adding primes to equal numbers in column 4, their remainders must add to 4. This only happens when both primes come from column 5, and never when one or both primes come from column 1. For example, 5 and 17 are both in column 5. They add to 22, an even number in column 4. On the other hand, if we add 5 to any prime in column 1 we get a resulting number in column 6, and never in column 4. Therefore, column 4 numbers are restricted just like those in column 2. Only numbers from column 6 use primes from both columns 1 and 5. If we were to create tables of larger numbers similar to Table 17, we would see that this impact on the Goldbach numbers continues with larger numbers. The sum of the three columns; 2, 4, and 6, for the next 20 rows of Goldbach numbers (from 122 to 240) are: column 2 = 127, column 4 = 146, and column 6 = 248. Therefore, column 6 continues to lead in the number of ways to add two primes to yield an even number.

From all the above we can now understand why the bands appear in Goldbach's Comet. The even numbers falling into certain columns in arrays can be written as sums of primes, using more of the primes than numbers from other columns. Such numbers will have larger Goldbach numbers and appear clustered together in a band in the comet. As a general rule we can say the largest Goldbach numbers will come from those numbers with the greatest number of different factors, while the smallest Goldbach numbers come from numbers which contain a 2 and one other large prime factor. From Table 15 we see the largest Goldbach number listed is 27 which comes from $390 = 2 \cdot 3 \cdot 5 \cdot 13$. A larger number with a significantly smaller Goldbach number is $398 = 2 \cdot 199$ whose Goldbach number is only 7.

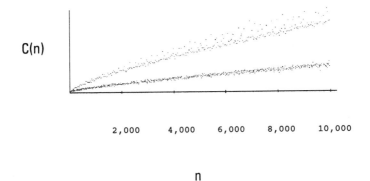

FIGURE 43. Goldbach's Comet for select even numbers. The lower band is produced by even numbers composed of the prime 2 plus one additional large prime. The upper band is produced by even numbers that include 2, 3, and 5 among their prime factors.

Figure 43 shows a Goldbach Comet for specific kinds of numbers. The lower band is produced by numbers composed of the prime 2 plus one additional large prime. On the other hand, the upper band is produced by numbers containing the factors 2, 3, and 5. Notice that the two bands are completely separated.

## ROUND NUMBERS

Since we have introduced the notion that some numbers are composed of numerous small primes, it is a good time to define *round numbers*. Round numbers are just those numbers that, when factored, contain a large number of primes, which may be different primes or one or more primes repeated. For example, $210 = 2 \cdot 3 \cdot 5 \cdot 7$ is considered a round number. But also considered round is $128 = 2 \cdot 2 \cdot 2 \cdot 2 \cdot 2 \cdot 2 \cdot 2 = 2^7$. Mathematicians have two ways for measuring the roundness of numbers. If we factor a number into its primes we can show it as:

$$N = P_1^{a_1} P_2^{a_2} P_3^{a_3} \ldots P_m^{a_m}$$

Each $P$ is a different prime, and the exponents $a_1, a_2, a_3, \ldots, a_m$ show how many times each different prime occurs in the number. Hence, $500 = 2 \cdot 2 \cdot 5 \cdot 5 \cdot 5$ or $2^2 \cdot 5^3$. One way to measure roundness is just to count the number of different primes and ignore how many times each is used. The number of different primes occurring in a number is $\omega(n)$. Hence, $\omega(500) = 2$ since there are two different primes in the number 500. A second way to measure roundness is to account for the number of times each prime appears by simply adding the exponents. The total number of primes occurring in a number is $\Omega(n)$. This means that $\Omega(500) = 5$, since a total of five primes are contained in 500, that is the prime 2 occurs twice and 5 occurs three times. Notice that to compute the 5 we simply added the exponents together for the primes of 2 and 5.

Now that we have these two measures of roundness in numbers, we might ask how fast they grow in size as integers get larger and larger. If $\omega(n)$ is large compared to $\omega(n-2)$, $\omega(n-1)$, $\omega(n+1)$, $\omega(n+2)$, or in general, if it is large compared to its neighboring values of $\omega(n)$, then we say that $n$ is round. Otherwise it is not.

Likewise, if $\Omega(n)$ is large compared to its neighboring values, then $n$ is considered round. The surprising result from studying how numbers factor is that round numbers are very rare.[8] The numbers we frequently encounter are often round. One thousand is round because $1000 = 2^3 \cdot 5^3$. One million is also round since $1{,}000{,}000 = 2^6 \cdot 5^6$. The frequency of round numbers we encounter in normal daily living would suggest they are quite numerous. But the opposite is the case. We can understand this by investigating how both $\Omega(n)$ and $\omega(n)$ increase as $n$ increases.

If we compute $\omega(n)$ and $\Omega(n)$ for various $n$, we see that their values tend to jump around in a haphazard manner, not unlike Goldbach numbers. Table 18 lists the first 100 values for both measures of roundness. Mathematicians have identified a function that is an approximate measure of $\omega(n)$ and $\Omega(n)$. While their values do jump around, there is a value we can compute that is almost always close to the individual values of $\omega(n)$ and $\Omega(n)$. This impreciseness is a strange way to hear mathematicians talk, for mathematics is generally considered to be the most precise activity carried out by humans. Yet, many mathematical phenomena behave in such an erratic manner that we must measure them through less than precise means. One way to do this is to talk of the *normal order* of functions. Normal order is defined as a measure that is approximate to the functions in almost all values. Hence, the measure is close except for a very few exceptions.

With this rather vague definition of normal order we present a wonderful theorem which measures the growth of both $\omega(n)$ and $\Omega(n)$.

*Theorem:* The normal order of $\omega(n)$ and $\Omega(n)$ is $\ln(\ln n)$

The definition of $\ln(\ln n)$ is simply to take the natural logarithm of $n$, and then take the natural logarithm of the answer. From this theorem we can deduce that most values of $\omega(n)$ and $\Omega(n)$ are approximately $\ln(\ln n)$. For example, if $n = 1{,}000{,}000$ then $\omega(n)$ and $\Omega(n)$ are both close to $\ln(\ln 1000000) = 2.63$. What this says is that the majority of numbers close to one million have between two and three factors! If we just take a moment to think about it, this is incredible. For example, $\Omega(1{,}000{,}000) = 12$, which says one million has a total of 12 factors, and is very round. But, on average, the numbers in that range have many fewer factors.

Table 19 lists the values of $\omega(n)$, $\Omega(n)$, and the factors of the 20 numbers between 999,991 and 1,000,010. At the bottom of the table we show the average values for both $\omega(n)$ and $\Omega(n)$. While a few numbers have a high degree of roundness, including 1,000,000, most are close to our estimated value of $ln\ (ln\ 1,000,000) = 2.62$. As numbers increase beyond one million, the value of $ln\ (ln\ n)$ becomes an even better measure of roundness. In Table 20 we show the same values for the 20 numbers between 999,999,991 and

**Table 18.**  First 100 Values of $\omega(n)$ and $\Omega(n)$

| $n$ | $\omega(n)$ | $\Omega(n)$ | $n$ | $\omega(n)$ | $\Omega(n)$ | $n$ | $\omega(n)$ | $\Omega(n)$ | $n$ | $\omega(n)$ | $\Omega(n)$ |
|---|---|---|---|---|---|---|---|---|---|---|---|
| 1 | 1 | 1 | 26 | 2 | 2 | 51 | 2 | 2 | 76 | 2 | 3 |
| 2 | 1 | 1 | 27 | 1 | 3 | 52 | 2 | 3 | 77 | 2 | 2 |
| 3 | 1 | 1 | 28 | 2 | 3 | 53 | 1 | 1 | 78 | 3 | 3 |
| 4 | 1 | 2 | 29 | 1 | 1 | 54 | 2 | 4 | 79 | 1 | 1 |
| 5 | 1 | 1 | 30 | 3 | 3 | 55 | 2 | 2 | 80 | 2 | 5 |
| 6 | 2 | 2 | 31 | 1 | 1 | 56 | 2 | 4 | 81 | 1 | 4 |
| 7 | 1 | 1 | 32 | 1 | 5 | 57 | 2 | 2 | 82 | 2 | 2 |
| 8 | 1 | 3 | 33 | 2 | 2 | 58 | 2 | 2 | 83 | 1 | 1 |
| 9 | 1 | 2 | 34 | 2 | 2 | 59 | 1 | 1 | 84 | 3 | 4 |
| 10 | 2 | 2 | 35 | 2 | 2 | 60 | 3 | 4 | 85 | 2 | 2 |
| 11 | 1 | 1 | 36 | 2 | 4 | 61 | 1 | 1 | 86 | 2 | 2 |
| 12 | 2 | 3 | 37 | 1 | 1 | 62 | 2 | 2 | 87 | 2 | 2 |
| 13 | 1 | 1 | 38 | 2 | 2 | 63 | 2 | 3 | 88 | 2 | 4 |
| 14 | 2 | 2 | 39 | 2 | 2 | 64 | 1 | 6 | 89 | 1 | 1 |
| 15 | 2 | 2 | 40 | 2 | 4 | 65 | 2 | 2 | 90 | 3 | 4 |
| 16 | 1 | 4 | 41 | 1 | 1 | 66 | 3 | 3 | 91 | 2 | 2 |
| 17 | 1 | 1 | 42 | 3 | 3 | 67 | 1 | 1 | 92 | 2 | 3 |
| 18 | 2 | 3 | 43 | 1 | 1 | 68 | 2 | 3 | 93 | 2 | 2 |
| 19 | 1 | 1 | 44 | 2 | 3 | 69 | 2 | 2 | 94 | 2 | 2 |
| 20 | 2 | 3 | 45 | 2 | 3 | 70 | 3 | 3 | 95 | 2 | 2 |
| 21 | 2 | 2 | 46 | 2 | 2 | 71 | 1 | 1 | 96 | 2 | 6 |
| 22 | 2 | 2 | 47 | 1 | 1 | 72 | 2 | 5 | 97 | 1 | 1 |
| 23 | 1 | 1 | 48 | 2 | 5 | 73 | 1 | 1 | 98 | 2 | 3 |
| 24 | 2 | 4 | 49 | 1 | 2 | 74 | 2 | 2 | 99 | 2 | 3 |
| 25 | 1 | 2 | 50 | 2 | 3 | 75 | 2 | 3 | 100 | 2 | 4 |

**Table 19.** Values of $\omega(n)$ and $\Omega(n)$
for $n$ between 999,991 and 1,000,010

| $n$ | $\omega(n)$ | $\Omega(n)$ |
|---|---|---|
| 999,991 | 3 | 3 |
| 999,992 | 3 | 6 |
| 999,993 | 2 | 2 |
| 999,994 | 3 | 3 |
| 999,995 | 2 | 2 |
| 999,996 | 4 | 5 |
| 999,997 | 2 | 2 |
| 999,998 | 3 | 4 |
| 999,999 | 5 | 7 |
| 1,000,000 | 2 | 12 |
| 1,000,001 | 2 | 2 |
| 1,000,002 | 3 | 3 |
| 1,000,003 | 1 | 1 |
| 1,000,004 | 3 | 5 |
| 1,000,005 | 4 | 4 |
| 1,000,006 | 3 | 3 |
| 1,000,007 | 2 | 2 |
| 1,000,008 | 5 | 8 |
| 1,000,009 | 2 | 2 |
| 1,000,010 | 4 | 4 |
| Average | 2.9 | 4.0 |

1,000,000,010. Reviewing the factors listed demonstrates that most numbers have few factors.

Remembering our graph of the growth of $ln\ n$, and how slow that growth was in comparison to $n$, we can appreciate that $ln\ (ln\ n)$ grows excruciatingly slower than $n$. If we consider the numbers in the range of $10^{80}$, which is the approximate number of protons in the universe, we can determine that most such numbers have around five factors since $ln\ (ln\ 10^{80}) = 5.22$. Modern public-key encryption methods use key numbers that are approximately 155 digits long or numbers in the range of $10^{155}$. Computing normal roundness for such large numbers we find $ln\ (ln\ 10^{155}) = 5.88$. This means that most such numbers contain approximately six factors. Of course, the fact that most large numbers contain few factors makes factoring such numbers even more difficult.

**Table 20.** Values of $\omega(n)$ and $\Omega(n)$
for $n$ between 999,999,991 and
1,000,000,010

| $n$ | $\omega(n)$ | $\Omega(n)$ |
|---|---|---|
| 999,999,991 | 2 | 2 |
| 999,999,992 | 3 | 5 |
| 999,999,993 | 3 | 3 |
| 999,999,994 | 3 | 3 |
| 999,999,995 | 4 | 4 |
| 999,999,996 | 3 | 4 |
| 999,999,997 | 3 | 3 |
| 999,999,998 | 3 | 3 |
| 999,999,999 | 3 | 6 |
| 1,000,000,000 | 2 | 18 |
| 1,000,000,001 | 5 | 5 |
| 1,000,000,002 | 5 | 5 |
| 1,000,000,003 | 3 | 3 |
| 1,000,000,004 | 3 | 5 |
| 1,000,000,005 | 3 | 3 |
| 1,000,000,006 | 2 | 2 |
| 1,000,000,007 | 1 | 1 |
| 1,000,000,008 | 5 | 9 |
| 1,000,000,009 | 1 | 1 |
| 1,000,000,010 | 4 | 4 |

## WARING'S PROBLEM

We have already mentioned several problems in additive number theory including the partition problem and Goldbach's Conjectures. Another famous such problem is due to Waring. Edward Waring (1734–1793) was an English mathematician who was educated at Cambridge, the same university where Hardy, Littlewood, and Ramanujan worked. The school gave a very difficult test, called the Tripos, to its mathematics students generally during the student's third year. Individual students were known by how well they performed on the exam. The highest scorer for any year was called senior wrangler, with the second highest scorer being second wrangler, then third wrangler, and so on. To achieve senior wrangler was like winning a gold medal at the mathematical olympics for any English mathematics student, and Waring was senior wrangler when he graduated. (As a side note, we might mention that G.H. Hardy, who took the Tripos his second year, was fourth wrangler, while his friend Littlewood was senior wrangler.[9] The Tripos is no longer given.)

Waring's work contained many important results, several of which were published in his book, *Meditationes algebraicae*, which came out in 1770. We have already mentioned Wilson's Theorem regarding prime numbers, named for Wilson by Waring. The first written record of Goldbach's Conjectures is found in Waring's *Meditationes algebraicae*. Waring may also have been the first to use the important ratio test to test for the convergence of infinite series, although it is now called Cauchy's test, after Augustin-Louis Cauchy (1789–1857). Unfortunately, the *Meditationes* was not widely read, and Waring did not receive the credit during his life that he should have.

In his book, Waring conjectured about integers being written as the sum of other integers raised to various powers. For example, we can write $13 = 9 + 4 = 3^2 + 2^2$. From this we see that 13 can be written as the sum of two squares. Can every number be written as the sum of two squares? No. For example, 12 cannot be written so. When we try, we get the following combinations:

$$12 = 1^2 + 11, 12 = 2^2 + 8, \text{ and } 12 = 3^2 + 3$$

The next square, $4^2$, is larger than 12. Hence, these three combinations use one square, but the second number is not a natural square.

Now comes the interesting question. What is the least number of squares needed to represent every positive integer? We know from our example with 12 that it cannot be two squares. Will three squares suffice? No. In fact, in 1770 Joseph-Louis Lagrange proved that every positive integer can be written as the sum of no more than four squares. We can write 12 as the sum of three squares: $12 = 2^2 + 2^2 + 2^2$, and any number, no matter how large, can be written using at most four squares.

Waring went beyond squares and conjectured that every positive integer can be written as the sum of no more than nine cubes, and no more than 19 fourth powers. Now we come to the general question lying at the heart of the Waring problem. Let's define $g(k)$ as the smallest number of $k$th powers needed to represent all positive integers. Hence, we know $g(2) = 4$, because 4 is the smallest number of squares needed to represent every positive integer. Waring guessed that $g(3) = 9$, or that it would take nine cubes (or fewer) to represent every integer. It is logical to assume that some very large numbers require a full nine cubes, but this is not the case. Consider the number 23. We cannot use $3^3$ (27) in the sum for 23 because it is too large. Therefore, 23 must be shown as the sum of ones and twos squared. Hence, the smallest representation of 23 as the sum of cubes is:

$$23 = 2^3 + 2^3 + 1^3 + 1^3 + 1^3 + 1^3 + 1^3 + 1^3 + 1^3$$

The only other number known to require nine cubes is 239:

$$239 = 5^3 + 3^3 + 3^3 + 3^3 + 2^3 + 2^3 + 2^3 + 2^3 + 1^3$$

If we try to write every integer as the sum of eight cubes, we will fail for only 23 and 239. Waring also guessed that $g(4) = 19$.

Now we can state Waring's problem. Given the power $k$, what is the least number of $k$th powers needed to represent every positive integer? Or, for any $k$, what is $g(k)$? This is a very hard problem, and has not, in its most general form, been solved. It really contains two problems: (1) Does $g(k)$ exist for every $k$; and (2), if so, what is it? As a general rule we can say that $k + 1 \le g(k)$. Therefore, if we try to represent all numbers as the sum of $k$ numbers to the $k$th power, we will fail. Yet, this is only a lower bound, and does not tell us what $g(k)$ should be for any particular $k$.

Although substantial work has been carried out on Waring's problem, only specific cases have been solved.[10] For $k = 2$ through 10 we have:

$g(2) = 4$   Conjectured by Fermat, proved by Lagrange, 1770
$g(3) = 9$   Conjectured by Waring, proved by Wieferich, 1912
$g(4) = 19$  Conjectured by Waring, proved by the team of
             Balasubramanian, Dress, Deshouillers, 1986
$g(5) = 37$  Proved by Chen, 1964
$g(6) = 73$  Proved by Pillai, 1940

$143 \leq g(7) \leq 3806$
$279 \leq g(8) \leq 36{,}119$
$548 \leq g(9) \leq$ ?
$1079 \leq g(10) \leq$ ?

From this list, we see that only the specific cases for $n = 2, 3, 4,$ 5, and 6 are solved. We have a little information for 7 through 10. But what about a general formula? If we put the known $g(n)$ in a sequence $(4, 9, 19, 37, 73, \ldots)$ we need some technique to easily get the next number.

Another part of the Waring problem is to ask: What is the smallest number of powers of $k$ for which there will be only a finite number of failures? For example, we know it takes four squares to represent all integers. But if we try to use only three squares? Will the number of integers that fail be only finite? In other words, is there some large number $n$ such that all larger numbers can be represented by only three squares? The answer is no. If we try to use only three squares to write numbers, there will be an infinite number of exceptions.

For another example we can look at cubes. We know that to represent all positive integers requires nine cubes. However, only two relatively small numbers need a full nine cubes. We now know that all larger numbers require only eight or fewer cubes. Hence, we know that it will require fewer than nine cubes to represent all positive integers beyond 239. Can we say the same for seven cubes? Does some integer exist such that seven cubes is enough to represent all larger numbers? Yes. Can we say the same about six? Now we run into problems.

We will let $G(k)$ be the smallest number of $k$th powers for which there are only a finite number of exceptions. Hence, if we try to represent numbers with fewer $k$th powers than $G(k)$, there will be an infinite number of failures. We know that $G(2) = 4$, just as $g(2) = 4$. While $g(3) = 9$, what is $G(3)$? We can only say that $4 \leq G(3) \leq 7$, or that $G(3)$ is some number between 4 and 7.

This second Waring problem seems to be the more difficult. Yet, it is of just as much interest to mathematicians as $g(k)$, because small numbers seem to have their own peculiarities which the large numbers do not share. This means some theorems only fail for small numbers and no others. For example, the proposition that all primes are odd numbers only fails for the number 2. Therefore, $G(k)$ (only finite failures) is just as important as $g(k)$ (no failures).

In general we can say that $k + 1 \le G(k) \le g(k)$. For specific $G(k)$ where $k = 2$ through 10 we know the following:

$G(2) = 4$

$4 \le G(3) \le 7$

$G(4) = 16$

$6 \le G(5) \le 21$

$9 \le G(6) \le 31$

$8 \le G(7) \le 45$

$32 \le G(8) \le 62$

$13 \le G(9) \le 82$

$12 \le G(10) \le 102$

We know $G(k)$ exactly only for $k = 2$ and 4. For other values we only know a range. Because of the difficulty of Waring's problem, it may be some time before we have a general solution that allows us to compute $g(k)$ and $G(k)$ directly from $k$. Yet, we do have some information regarding a general solution.

In 1909 the brilliant German mathematician David Hilbert (1862–1943) proved that for any $k$, $g(k)$ existed, and was finite. If $g(k)$ exists, then we know that $G(k)$ exists, for the worst case will be when $G(k) = g(k)$. There have also been efforts to find some upper bounds on $g(k)$ and $G(k)$, depending on $k$. In 1954 G.J. Rieger showed that $g(k) \le 2^{32(k+1)!}$ and in 1984 R. Balasubramanian and C.J. Mozzochi showed that:

$$G(k) \le \frac{-2ln(3k) - ln(6k)}{ln\left(\dfrac{k-1}{k}\right)} - 4$$

The above equations produce rather larger upper bounds for $g(k)$ and $G(k)$. For example, if $k = 3$ then we get the upper bounds of $g(3) \le 2^{758}$ (a rather larger number) and $G(3) \le 14$, a much more reasonable bound than for $g(k)$, but considerably larger than the bound of 7 we currently have.

And now, one last wrinkle to additive problems. A new problem is possible if we combine the Waring problem with the Goldbach Conjecture. Consider writing numbers as sums of $k$th powers, but only when the numbers used in the sums are primes. When $k = 1$ we get the Goldbach Conjecture. That is, is every sufficiently large number represented by the sum of two or three primes raised to the first power? In 1938 Vinogradov showed that for every $k$, there was an integer, $V(k)$, so that every sufficiently large number was the sum of $V(k)$ primes all raised to the $k$th power. As an example, we have $40 = 5^2 + 3^2 + 2^2$, where $k = 2$, and each number that is raised to the $k$th power is a prime. It is obvious that all numbers can't be written this way. However, Vinogradov proved that if we let $n$ be large enough, then every number bigger than $n$ can be written as the sum of squares of primes. The problem is to find the $V(2)$ or how many squares we need. This is called the Waring–Goldbach Problem.

In this chapter we have concentrated on additive problems including Goldbach's Conjecture and Waring's problem. The original questions posed are very simple and would suggest quick answers, but the answers are difficult and slow in coming. Why is that? Why can't we just look at the questions being asked and determine which will be hard to answer and which will be easy? When studying the natural number sequence we are constantly confronted by this dilemma. The question of whether an infinity of primes exist was easily answered by Euclid over 2000 years ago, and the average school-age child can be taught the proof in a few minutes. Yet, Goldbach's simple question is still unanswered, even after the giants of mathematics have tried their hand at it. This teaches us that deep questions need not come from complex or convoluted problems. They can come from innocent questions asked by the average man or woman (or child) on the street. This characteristic of mathematics adds to its charm while suggesting a broader question: How do we recognize the deep and hard questions from the easy? Maybe the answer to this is very deep.

## CHAPTER THIRTEEN

# DEEPEST MYSTERIES

*If I were to awaken after having slept for a thousand years, my first question would be: Has the Riemann hypothesis been proven?*[1]

DAVID HILBERT (1862–1943)

*W*e have left the best for last. We say it is best for several reasons. First, the mysteries we are about to plumb regarding the natural number sequence are the deepest and most magical. Second, understanding them involves the greatest problem in all of mathematics. We are going to study the Riemann hypothesis, a hypothesis that everyone assumes to be true, but nobody can prove. The mathematician quoted above, David Hilbert, was one of the greatest mathematicians to bridge the 19th and 20th centuries. At a 1900 Paris conference he proposed 23 mathematical problems that should occupy the attention of mathematicians during the 20th century. About half have now been solved. One of the 23 still unsolved is the Riemann hypothesis.

One reason the Riemann hypothesis is not better known is that it's basic formulation is more complex and requires considerably more knowledge of number theory. However, with the previous pages of this volume, we have prepared ourselves for Riemann. Before we sit down to the main course though, we will whet our appetites with some tantalizing mathematical hors d'oeuvres.

## COMPLEX NUMBERS

Up to this point we have restricted our attention to the numbers that make up the real number line: integers, fractions, algebraic numbers, and transcendental numbers. Another kind of number was only mentioned. Remember when we looked at Euler's wonderful formula, $e^{\pi\sqrt{-1}} + 1 = 0$? We encountered the strange symbol, $\sqrt{-1}$, and pointed out that there is no real number that, when multiplied by itself, gives us a value of –1. To understand the need for such numbers, consider the simple equation $X^2 + 1 = 0$. The square of any real number is always positive, and adding 1 to a positive number will never give us a zero. Hence, we must look elsewhere for our solution.

To find solutions to such equations we define the *complex number field*. Simply take two straight lines and set them at right angles to each other, as in Figure 44. These two lines are called the axes: the horizontal (or real) axis and the vertical (or imaginary) axis. The point where they intersect is called the origin, which we designate as the *complex number (0,0)*. On a number line, every point is

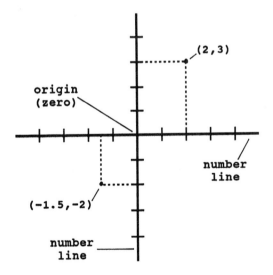

FIGURE 44. The complex number plane showing the complex numbers (2,3) and (–1.5,–2). Such numbers can also be written in the form 2 + 3i and –1.5 – 2i.

associated with a real number. However, in the two-dimensional plane in Figure 44 every point of the entire plane is associated with a *complex number*. We identify each point on the plane (complex number) by using two unique real numbers. In Figure 44 we have identified the two complex numbers (2,3) and (–1.5,–2). The first real number in a complex number pair represents the complex number's distance from the vertical line, and is called the real part. The second number of the pair is the distance from the horizontal line; it is called the imaginary part. Usually, instead of writing complex numbers as $(a,b)$, we write them as $a + bi$, where the $i$ stands for the imaginary unit, $\sqrt{-1}$. Writing complex numbers in this form allows us to manipulate them in computations.

We can now see that the real numbers are just a subset of the complex numbers, for when the imaginary part of a complex number is equal to zero, then the corresponding point is located on the horizonal line of Figure 44. Why do we need these strange new numbers? Much of modern mathematics is based on complex numbers, and they are used extensively in the sciences. And now we are ready to use them in understanding the Riemann hypothesis.

## CHASING AFTER $\pi(n)$

The great body of mathematics is constructed upon the concept of the natural number sequence. The foundation of the natural number sequence is the set of all prime numbers, for prime numbers multiply together to form every natural number. In other words, if we had only the prime numbers we could generate every other natural number simply by multiplying the various primes together. Thus, by understanding the distribution of prime numbers, we gain control of the number sequence itself. This is why the study of prime numbers is so important to the foundation of mathematics.

The first question regarding primes (How many are there?) was answered by Euclid of ancient Greece. The next question (How are they distributed?) has been perplexing mathematicians to this very day. We cannot determine the distribution of primes on the small scale because they occur in such a random fashion. For example, we can't easily identify the next prime occurring after a certain number $n$ (there's no formula or algorithm). We also find it very difficult to determine if many large numbers are prime or composite. However, we can see some regularity in the overall flow of the primes. One basic question is: Given any number $n$, how many primes exist that are less than $n$? As we've pointed out, this number is defined as $\pi(n)$, called the *prime counting function*. To compute $\pi(n)$ exactly we must laboriously count all the primes from 2 through $n$, a horrendous project when $n$ is large. The largest computed value of $\pi(n)$ is for $n = 10^{17}$ or $\pi(10^{17}) = 2,625,557,157,654,233$, calculated by M. Deleglise in 1992.[2] What we really need is a formula for computing $\pi(n)$ exactly. But the best that is possible is a formula that approximates $\pi(n)$, and thus captures the essence of prime distribution. It is this long, hard road to unraveling $\pi(n)$ that eventually leads to the Riemann hypothesis.

One of the first to seriously attack the problem of understanding $\pi(n)$ was the French mathematician, Adrien-Marie Legendre (1752–1833), who gave an estimate to $\pi(n)$ in 1798:

$$\pi(n) \approx n/(ln\ n - 1.08366)$$

where $ln\ n$ is the natural logarithm of $n$. While this formula isn't too bad, it breaks down for large $n$. However, the formula is remarkable because it establishes a link between prime numbers and the natural logarithms of numbers, which in turn is defined in terms of the constant $e$, the limit to continuously compounded interest.

Next, in 1792, Carl Gauss, at the age of 15, noticed that the primes increased as a function of $n/(ln\ n)$, which is close to Legendre's claim. However, it was Gauss' formula that proved the better of the two and has become known as the *Prime Number Theorem*. What Gauss suspected, but could not prove, was that as $n$ grows larger and larger, $n/(ln\ n)$ becomes an ever better estimate of $\pi(n)$. Another way of expressing this idea symbolically is:

$$\lim_{n \to \infty} \pi(n) = \frac{n}{ln\ n}$$

As $n$ gets larger and larger, $n/(ln\ n)$ gets closer to the true value of $\pi(n)$.

As previously stated, Gauss also proposed another formula called the *logarithmic integral* of $n$, $li(n)$, which is based on integral calculus. Gauss reasoned that if $n/(ln\ n)$ is an approximation of the number of primes less than $n$, then $1/(ln\ n)$ is an approximation of the probability that $n$ is a prime. The sum of all such probabilities from 2 through $n$ becomes another estimate of the number of primes less than $n$. This can be approximated with his integral, $li(n)$. After a sufficiently large $n$, $li(n)$ always gives a better estimate to $\pi(n)$ than $n/(ln\ n)$, yet both are based on the same logarithmic relationship.

The proof that $n/(ln\ n)$ and $li(n)$ give closer and closer approximations to $\pi(n)$ had to wait until 1896 when two mathematicians, the Frenchman Jacques Hadamard and the Belgian C.J. de la Vallée-Poussin, independently proved the theorem.[3] How did Hadamard and Vallée-Poussin prove the Prime Number Theorem? They used a function called the Riemann zeta function. We're getting closer!

## ZETA

In 1737 Euler made one of the most remarkable discoveries in mathematics when he noticed the following wonderful identity:

$$\sum_{n=1}^{\infty} \frac{1}{n^s} = \prod_{p = primes} \frac{1}{1 - \dfrac{1}{p^s}}$$

In order to fully appreciate this relationship we show both sides in the expanded form.

$$\frac{1}{1^s} + \frac{1}{2^s} + \frac{1}{3^s} + \frac{1}{4^s} + \ldots = \left(\frac{1}{1 - \dfrac{1}{2^s}}\right)\left(\frac{1}{1 - \dfrac{1}{3^s}}\right)\left(\frac{1}{1 - \dfrac{1}{5^s}}\right)\ldots$$

On the left we have the infinite series running through all the positive integers, while on the right, the product runs through all primes, $p$. For Euler, the $s$ on both sides was a variable that could take the value of any real number.

Euler's identity expresses a fundamental relationship between the sequence of natural numbers and the primes. What is remarkable is not only that Euler discovered this relationship, but that he and future mathematicians recognized just how important it was in cementing the link between whole numbers and primes. That Euler's identity is true is not obvious even to most mathematicians, so it is worth the effort to satisfy ourselves that it is correct.

First we consider the following identity:

$$\left(\frac{1}{1-\frac{1}{p^s}}\right) = 1 + \frac{1}{p^s} + \frac{1}{p^{2s}} + \frac{1}{p^{3s}} + \frac{1}{p^{4s}} + \dots$$

This identity can be verified by using long division on the left side of the equation to arrive at the form on the right side. Using the above identity, we replace each of the terms on the right side of Euler's identity with the corresponding infinite series:

$$\sum_{n=1}^{\infty} \frac{1}{n^s} = \left(1 + \frac{1}{2^s} + \frac{1}{2^{2s}} + \frac{1}{2^{3s}} + \dots\right)\left(1 + \frac{1}{3^s} + \frac{1}{3^{2s}} + \frac{1}{3^{3s}} + \dots\right)\dots$$

Now, if we multiply every one of the infinite series (that means multiplying every term in each set of parentheses with every term in every other set of parentheses) a miraculous thing happens: We get every possible combination of prime numbers in the denominators, but each combination occurs only once. This means that when multiplied out, the right side becomes identical to the left side. Hence, we have $1/n^s$ where $n$ runs through all whole numbers.

This wonderful series, $\Sigma 1/n^s$, which is equal to the product of the terms involving the primes, is called the zeta function and is designated as $\zeta(s)$, or:

$$\zeta(s) = \sum_{n=1}^{\infty} \frac{1}{n^s} = \prod_{p} \left(\frac{1}{1-\frac{1}{p^s}}\right)$$

The zeta function has been studied in great detail as mathematicians attempt to unravel the secret connections between the natural

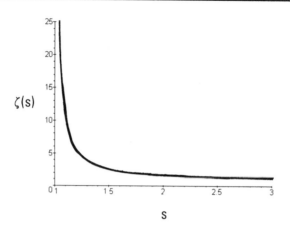

FIGURE 45. The value of the zeta function, $\zeta(s)$, for $s > 1$.

number sequence and the primes. For some values of $s$, $\zeta(s)$ diverges and for others it converges. For $s = 1$, the zeta function is just the harmonic series, and we know that the harmonic series diverges. If we let $s$ be any number larger than 1 ($s > 1$) then the zeta function converges to a limit. Figure 45 is a graph of the zeta function where $s > 1$. Notice that as $s$ increases, the value of $\zeta(s)$ decreases since the denominators of the fractions we are adding together in the infinite series are growing larger.

## RIEMANN

Georg Friedrich Bernhard Riemann (Figure 46) was born September 17, 1826 in the little village of Breselenz, near Hanover, Germany. His father was a Lutheran pastor who had fought in the Napoleonic Wars, before settling in Breselenz with his wife, Charlotte. She died while their six children were still young, and the father was left to raise them on his meager income.[4]

Georg, their second child, was frail and shy. Yet, his father was determined his son should follow him in the ministry. When Georg was ten he began his instruction in arithmetic and geometry from Schulz, a local teacher. At 14, he went to live with his grandmother in Hanover, and entered the local gymnasium. In 1842, two years

FIGURE 46. Georg Friedrich Riemann, 1826–1866.

after his arrival, Georg's grandmother died. Georg then entered the Luneberg Gymnasium where he stayed until 1846. At 19, he entered the University of Göttingen, majoring in philology and theology. By this time he was aware that his personality and disposition (especially his fear of public speaking) were not suited for preaching, although he was, and would remain, a devout Christian. His real talent was in mathematics. Receiving permission from his father, Georg formally changed his course of study, and in his second year moved to the University of Berlin. In 1949 he returned to Göttingen where he became a student of Carl Gauss, among others, and finished his doctorate. By 1854 he was a lecturer at Göttingen, and three years later became a professor of mathematics there. Not only did Riemann study mathematics, he also devoted much of his time to the physical sciences, especially physics.

Riemann was a brilliant mathematician. Carl Gauss, his teacher, was certainly the greatest mathematician of the 19th century, and maybe of all time. Yet, Riemann's talent has been favorably compared to his teacher's.[5] Riemann is best known for his contributions in geometry where he was one of the cofounders of non-Euclidean geometry. He modernized geometry by demonstrating that space could be characterized, not as a cauldron of lines and points, but as sets of $n$-tuples of numbers combined according to different sets of rules. The metric of space is defined by a formula for computing the distance between two points ($n$-tuples), and the rules for computing this distance could be changed within certain bounds to yield different metric spaces.

Riemann's contribution to number theory was an eight-page memoir entitled, "On the Number of Prime Numbers Under a Given Magnitude," which was printed in the November 1858 notice of the Berlin Academy. In this memoir Riemann showed how to extend the zeta function, $\zeta(s)$, to cover not only real numbers, but complex numbers as well. He did this by extending the function to all complex numbers, $s = a + bi$, including values less than 1. To do this, he used analytical continuation, a technique from classical complex analysis. This new extended zeta function became Riemann's zeta function. For values of $a > 1$ and $b = 0$, the Riemann zeta function was the same as Euler's zeta function. At values $a \leq 1$, $\zeta(s)$ becomes more sophisticated. The mathematics of Riemann's extended function go beyond what we have covered. What is important to realize is that Riemann was able to extend Euler's zeta function to complex numbers, and in so doing, he evolved a much more powerful mathematical machine for analyzing the distribution of prime numbers.

Riemann, along with Euler, Legendre, Gauss, and others, wondered whether $n/(\ln n)$ was always an accurate function to estimate $\pi(n)$, the prime counting function, and if so, how closely did $n/(\ln n)$ approach $\pi(n)$ for large $n$? Now Riemann had a new approximation for $\pi(n)$, namely $li(n)$. How much better was $li(n)$ for approximating $\pi(n)$ than the older function $n/(\ln n)$? However, Riemann did not stop here. He constructed a new function that was

**Table 21.** Percent Error between $\pi(n)$ and the Three Functions: $n/ln(n)$, $li(n)$, and $R(n)$

| $n$ | $n/ln(n)$ | $li(n)$ | $R(n)$ |
|---|---|---|---|
| $10^8$ | 5.78 | $1.31 \times 10^{-4}$ | $1.68 \times 10^{-5}$ |
| $10^9$ | 5.10 | $3.35 \times 10^{-5}$ | $1.55 \times 10^{-6}$ |
| $10^{10}$ | 4.56 | $6.82 \times 10^{-6}$ | $4.02 \times 10^{-6}$ |
| $10^{11}$ | 4.13 | $2.81 \times 10^{-6}$ | $5.63 \times 10^{-7}$ |
| $10^{12}$ | 3.77 | $1.02 \times 10^{-6}$ | $3.92 \times 10^{-8}$ |
| $10^{13}$ | 3.47 | $3.15 \times 10^{-7}$ | $1.67 \times 10^{-8}$ |
| $10^{14}$ | 3.21 | $9.83 \times 10^{-8}$ | $5.99 \times 10^{-9}$ |
| $10^{15}$ | 2.99 | $3.52 \times 10^{-8}$ | $2.45 \times 10^{-9}$ |
| $10^{16}$ | 2.79 | $1.15 \times 10^{-8}$ | $1.17 \times 10^{-9}$ |

an improvement over $li(n)$, now called the Riemann function and designated as $R(n)$. How much better is $R(n)$ than $li(n)$ in estimating the number of primes less than $n$? Table 21 lists the percent error between $\pi(n)$ and the three functions used to estimate $\pi(n)$: $n/ln(n)$, $li(n)$, and $R(n)$. From Table 21 we see that $R(n)$ is a substantial improvement over the other two functions.

Then Riemann went one step further. He gave a formula for the exact difference between $\pi(n)$ and his new function $R(n)$. This is expressed by:

$$\pi(n) = R(n) - \sum_{\rho}^{\infty} R(n^\rho)$$

To make any sense of this equation, we must come to understand what $\rho$ (Greek rho) stands for when used in the equation. To do this, we must return to Riemann's zeta function.

The Riemann zeta function, $\zeta(s)$, has two kinds of zeros, that is, values of $s$ (which can be complex numbers of the form $s = a + bi$) for which the zeta function has a value of zero. For example, a specific complex number is $0.5 + 14.135i$. In this complex number, the real part is 0.5 and the imaginary part is 14.135. If we replace $s$ in the Riemann zeta function with this complex number, then the value of the zeta function becomes zero. Hence, $0.5 + 14.135i$ is called a zero of the zeta function, and $\zeta(0.5 + 14.135i) = 0$. Now the question becomes: Where on the complex plane are all the zeros of Riemann's zeta function, or for what complex numbers does the zeta function become zero?

It turns out that the zeta function is zero for all numbers, even negative real numbers, i.e., $\zeta(-2) = 0$, $\zeta(-4) = 0$, $\zeta(-6) = 0$. Yet, such zeros are of little interest to us, and are therefore called trivial zeros by mathematicians. A second set of zeros for $\zeta(s)$ are located in an area of the complex plane called the critical strip. The critical strip is that vertical strip between zero and 1 on the horizontal axis (see Figure 47).

The line that halves the critical strip consists of all those complex numbers whose real part is equal to $1/2$, or all $s$ such that $s = 1/2 + bi$, where $b$ can take any real number value. Riemann conjectured

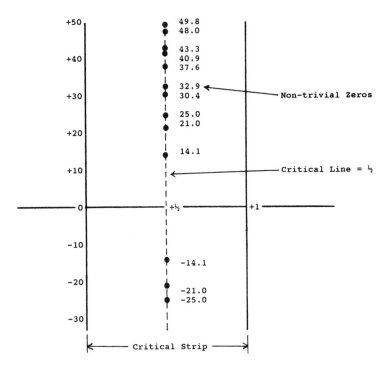

FIGURE 47. The complex number plane showing the critical strip and the location of various nontrivial zeros of the zeta function.

that all the zeros of the zeta function, except those that were trivial, would lie on this line in the middle of the critical strip. These zeros are called the nontrivial zeros of zeta. This is the Riemann hypothesis: *All non-trivial zeros of the Riemann zeta function fall on the line s = 1/2 + bi*, which is called the critical line. If even one zero of $\zeta(s)$ is within the critical strip, but off that line, then the Riemann hypothesis is false. No one has proved this hypothesis yet, and no one has found a single exception.

Now that we have defined the nontrivial zeros of the zeta function, we return to Riemann's exact expression for $\pi(n)$:

$$\pi(n) = R(n) - \sum_{\rho}^{\infty} R(n^{\rho})$$

What in the world are the rhos ($\rho$) in the above equation? They are the nontrivial zeros in the critical strip! Therefore, Riemann's exact expression for $\pi(n)$ is his Riemann function corrected by subtracting the sum of all the values of the Riemann function evaluated at *n* raised to the power of the *zeta's nontrivial zeros*! Hence, all nontrivial zeros of the zeta function define the exact correction needed on $R(n)$. This shows the deep connection between the prime counting function $\pi(n)$, Riemann's function, $R(n)$, and the zeros of the zeta function.

Knowing that the zeros in the critical strip define the correction between $\pi(n)$ and $R(n)$ does not help to know how many such zeros there are, or where they are. G.H. Hardy loved the zeta function, and proved in 1914 that the zeta function has an infinity of zeros on the critical line within the critical strip. Yet, this is still short of proving the Riemann hypothesis, because there could still be zeros within the critical strip that are off the critical line.

If the Riemann hypothesis is true, and all the nontrivial zeros are on the critical line, then the values of $\rho$ that correct Riemann's function all have the real number part equal to 1/2. This implies that the correction to $R(n)$ is much more orderly than if values of $\rho$ are off that line.

Mathematicians have computed nontrivial zeros in an attempt to find one that is off the critical line. To date, none have been found:

All *known* zeros of $\zeta(s)$ do fall on the critical line. The first 20 nontrivial zeros for $\zeta(s)$ are listed in Table 22. Remember that each one has a real part equal to $1/2$, which means all 20 fall on the critical line, just as the Riemann hypothesis predicts. The values listed in Table 22 are the $b$ values of $s = 1/2 + bi$.

If the zeros are all on the critical line, then the error in using $R(n)$ to estimate $\pi(n)$ for large values of $n$ is much smaller than if zeros exist off the critical line. That all zeros are on the critical line implies that the prime numbers are more orderly than we can prove today. In other words, if the Riemann hypothesis is false, then the position of the primes within the natural number sequence is more random than we suspect. In fact, we can make a stronger statement: The location of the zeros of the zeta function within the critical strip is equivalent to the problem of the location of prime numbers within the number sequence. If we can somehow prove the Riemann hypothesis, then it is possible to deduce much more precise theorems about the distribution of prime numbers. The proof of the Riemann hypothesis would have a far reaching impact on mathematics since many other theorems could be deduced if it were true.

All in all, Riemann made six conjectures regarding his extended zeta function in 1859, and five of the six proved to be true. It is only the sixth, the Riemann hypothesis, that has evaded all attempts at

**Table 22.** First 20 Nontrivial Zeros of the Zeta Function

| $n$ | $bi$ | $n$ | $bi$ |
|-----|--------|-----|--------|
| 1 | 14.135 | 11 | 52.970 |
| 2 | 21.022 | 12 | 56.446 |
| 3 | 25.011 | 13 | 59.347 |
| 4 | 30.425 | 14 | 60.833 |
| 5 | 32.935 | 15 | 65.113 |
| 6 | 37.586 | 16 | 67.080 |
| 7 | 40.919 | 17 | 69.546 |
| 8 | 43.327 | 18 | 72.067 |
| 9 | 48.005 | 19 | 75.705 |
| 10 | 49.774 | 20 | 77.145 |

a proof. After his eight-page memoir, Riemann did not return to a serious study of his function. As an adult his health was not good, and he suffered a nervous breakdown in 1851. At 36, he married Elise Koch, but one month later fell sick with pleurisy. From that time on he suffered from tuberculosis, tragically dying from it on July 20, 1866 when he was 39.

Riemann's life has a sad similarity to our good friend, Ramanujan, who also grew up in poverty and died at the young age of 32. How much more could Riemann have contributed had he not contracted an illness that today could have been controlled for years? Riemann's total contributions in mathematics would add up to just a single volume, but that volume of number theory, geometry, and mathematical physics was all pure genius. His impact on mathematics is almost incalculable.

## SKEWES' NUMBER

When we talked about large numbers, we mentioned Skewes' number which is defined as:

$$e^{e^{e^{79}}} \approx 10^{10^{10^{34}}}$$

We also learned that the prime counting function $\pi(n)$ is connected to the zeta function. As it turns out, Skewes' number and the prime counting function are also intimately connected.

Gauss constructed the logarithmic integral function, $li(n)$, to get a better estimate for $\pi(n)$:

$$\pi(n) \approx li(n) = \int_2^n \frac{1}{ln\ x}\ dx$$

Whenever we compute $li(n)$ for relatively small $n$s and compare it to $\pi(n)$, we discover that the estimate $li(n)$ is actually bigger than $\pi(n)$. If we let $D(n)$ be the difference between the two, then we have: $D(n) = li(n) - \pi(n)$, and $D(n)$ appears to always be positive and increasing. Hence, $li(n)$ tells us there are more primes less than $n$ than there really are. In fact, when we compute larger and larger $li(n)$ and compare these values to the correct $\pi(n)$, the absolute differences get larger but the percent differences get closer to zero.

In other words, $D(n)$ *appears* to keep growing as $n$ increases, but the percent error, $D(n)/\pi(n)$, goes to zero. While $li(n)$ becomes a better estimate of $\pi(n)$ as $n$ gets big, it looks like the difference between $li(n)$ and $\pi(n)$ just keeps increasing, and would tend to infinity as $n$ goes to infinity.

While everyone was assuming that $D(n)$ was always positive and growing, John Littlewood proved in 1914 that $D(n)$ not only decreases for sufficiently large $n$, but also becomes negative. Hence, at some very large number $n$, $\pi(n)$ would be bigger than $li(n)$. In fact, he proved something much stronger: $D(n)$ changes signs between positive and negative an infinite number of times.

Now everyone wanted to know for what $n$ did $D(n)$ become negative. Was it a somewhat big number or a really big number? S. Skewes, one of Littlewood's students, decided to tackle the problem. In his first assault on it in 1933, he had to assume that the unproven Riemann hypothesis was true. If so, then $D(n)$ would become negative before $n$ grew to:

$$e^{e^{e^{79}}} \approx 10^{10^{10^{34}}}$$

Now we know where Skewes' number comes from, i.e., it's a value of $n$ beyond which $D(n)$ has become negative.[6]

However, the problem with Skewes' number was that it depended on the Riemann hypothesis being true, and the number, itself, was too large to test, i.e., it was impossible to compute $\pi(n)$ for that size of number. Skewes went back to work. In 1937 he was able to come up with a new Skewes' number that did not depend upon the Riemann hypothesis being true. The new number was even bigger than the original one:

$$\textit{Skewes' number (1937)} = e^{e^{e^{e^{67}}}} \approx 10^{10^{10^{10^{29}}}}$$

In fact, this new Skewes' number is gigantic compared to the old one. Computing $\pi(n)$ and comparing it to $li(n)$ for this number was simply ridiculous. Nothing much changed for a number of years, even though a number of mathematicians worked on the problem. Finally, in 1955 Skewes came up with another, much improved, estimate:

$$Skewes'\ number\ (1955) = e^{e^{e^{e^{7.7}}}} \approx 10^{10^{10^{1000}}}$$

This is much better than the 1937 number but still much bigger than the original Skewes' number that assumed the Riemann hypothesis.[7] Skewes' had done his part, and now others pitched in. In 1966, R.S. Lehman got Skewes' number down to $1.65 \times 10^{1165}$. Then, in 1986, H.J.J. te Riele[8] managed to prove that Skewes' number was no bigger than $6.69 \times 10^{370}$. Finally, we were getting into the range of numbers we might someday attack through direct computation. However, we still don't know the exact value of $n$ at which $\pi(n)$ is bigger than $li(n)$.

## A CLOSE CALL

For a while it looked as if mathematicians were going to prove the Riemann hypothesis by proving a different conjecture, the Mertens' conjecture, and showing that it implied the Riemann hypothesis. To fully understand the conjecture we must first define a strange and wonderful function called the Möbius function, $\mu(n)$, named after August Ferdinand Möbius (1790–1868), the same mathematician who brought us the famous Möbius strip. August was also a student of Gauss (considered by Gauss to be his most talented student), who became a professor of astronomy and the director of the observatory at Leipzig in 1821.[9]

The fundamental theorem of arithmetic tells us that every positive integer factors into a unique set of prime numbers. We can show this symbolically as:

$$n = p_1^a \cdot p_2^b \cdot p_3^c \cdots$$

where $n$ is a positive integer and the various $p_1$ are primes. The exponents $(a, b, c, \ldots)$ show how many times each prime occurs. In 1832 Möbius defined the Möbius function in the following manner:

$\mu(n) = 0$   if any prime in the factorization of $n$ has an exponent of 2 or more,

$\mu(n) = 1$ if the number of different primes in the factorization
of $n$ is even,

$\mu(n) = -1$ if the number of different primes in the factorization
of $n$ is odd

Hence, the Möbius function has a value of 1 or $-1$ only if no prime
is repeated. For example, $\mu(4) = 0$ since $4 = 2^2$. However, $\mu(6) = 1$
because $6 = 2 \cdot 3$ (two primes), while $\mu(30) = -1$ because $30 = 2 \cdot 3 \cdot 5$
(three primes). For consistency, $\mu(1)$ is defined as 1. Table 23 shows
the Möbius value for the first 40 values of $n$. On the surface, it would
seem that the Möbius function is a curiosity, and unrelated to
anything deep in our study of primes. However, this is not the case.
For example, it is known that the Möbius function is intimately
related to the reciprocal of the zeta function and the product of
primes by the following wonderful formula:

$$\sum_{n=1}^{\infty} \frac{\mu(n)}{n^s} = \frac{1}{\zeta(s)} = \prod_{p=primes} \left(1 - \frac{1}{p^s}\right)$$

where $s$ is a complex number with real part greater then 1, and the
product on the right is over all primes. To fully appreciate the above
relation, we should see both the right and left side in expanded
form. In this example we will let $s = 1$.

**Table 23.** First 20 Values for the Möbius and Mertens Functions

| $n$ | Möbius $\mu(n)$ | Mertens $M(n)$ | $n$ | Möbius $\mu(n)$ | Mertens $M(n)$ |
|---|---|---|---|---|---|
| 1 | 1 | 1 | 11 | -1 | -2 |
| 2 | -1 | 0 | 12 | 0 | -2 |
| 3 | -1 | -1 | 13 | -1 | -3 |
| 4 | 0 | -1 | 14 | 1 | -2 |
| 5 | -1 | -2 | 15 | 1 | -1 |
| 6 | 1 | -1 | 16 | 0 | -1 |
| 7 | -1 | -2 | 17 | -1 | -2 |
| 8 | 0 | -2 | 18 | 0 | -2 |
| 9 | 0 | -2 | 19 | -1 | -3 |
| 10 | 1 | -1 | 20 | 0 | -3 |

$$\frac{\mu(1)}{1} + \frac{\mu(2)}{2} + \frac{\mu(3)}{3} + \frac{\mu(4)}{4} + \frac{\mu(5)}{5} + \cdots = \frac{1 \cdot 2 \cdot 4 \cdot 6 \cdot 10 \cdots \ldots}{2 \cdot 3 \cdot 5 \cdot 7 \cdot 11 \cdots \ldots}$$

Now we evaluate each $\mu(n)$ and simplify.

$$\frac{1}{1} - \frac{1}{2} + \frac{1}{3} - \frac{1}{5} + \frac{1}{6} - \frac{1}{7} + \frac{1}{10} + \cdots = \frac{1 \cdot 2 \cdot 4 \cdot 6 \cdot 10 \cdots \cdots}{2 \cdot 3 \cdot 5 \cdot 7 \cdot 11 \cdots \cdots}$$

The zeta function is unbounded when $s = 1$, hence we can conclude that the reciprocal of the zeta function is zero when $s = 1$. Therefore, when $s = 1$, both sides of the above equation are equal to zero.

If we add up all the values of $\mu(n)$ for 1 through $n$, then we get the Mertens function or:

$$M(n) = \sum_{j \leq n} \mu(j)$$

Therefore, if $\mu(1) = 1$, $\mu(2) = -1$, and $\mu(3) = -1$ then:

$$M(1) = \mu(1) = 1$$

$$M(2) = \mu(1) + \mu(2) = 1 + (-1) = 0$$

$$M(3) = \mu(1) + \mu(2) + \mu(3) = 1 + (-1) + (-1) = -1$$

Table 23 also gives the value of $M(n)$ for $n$ equal to 1 to 40. Notice that when $n$ is greater than 1, $M(n)$ is not positive. One interesting question is whether $M(n)$ ever becomes positive again.

The values of $\mu(n)$ are quite random and therefore difficult to predict. What about the value of $M(n)$? In 1897, Franz Mertens calculated a table of values[10] for both $\mu(n)$ and $M(n)$ that was 50 pages long and included values for $n$ up to 10,000. Studying the values of $M(n)$ compared to $n$, he conjectured that the absolute value of $M(n)$ (that is ignoring the negative sign in front) was always less than the square root of $n$. This became Mertens' conjecture: $|M(n)| < \sqrt{n}$ for all values of $n$. Just looking at Table 23 seems to support the conjecture, for by the time $M(n)$ reaches $-2$, $n$ is 5, and by the time $M(n)$ is $-3$, $n$ is 13.

In fact, once $n$ is greater than 200, it appears that a much stronger conjecture is possible. In 1897, R.D. von Sterneck conjectured that $|M(n)| < \frac{1}{2}\sqrt{n}$, because he had computed $M(n)$ for $n$ running up to five million, and found his conjecture still true. However, in 1960 W.B. Jurkat proved von Sterneck's conjecture false by proving a number existed that violated the conjecture. Soon after, H. Cohen[11] showed that 7,725,038,629 was the smallest integer for which $M(n) > \frac{1}{2}\sqrt{n}$.

Why all this fuss about Mertens' conjecture? It turns out that if Mertens' conjecture is true, then the Riemann hypothesis is also true. We use a finite sum of the Möbius function to define the Mertens function, $M(n)$. We use an infinite sum involving the Möbius function to define Riemann's function, $R(n)$ or:

$$R(n) = \sum_{m=1}^{\infty} \frac{\mu(m)}{m} \, li\,(n^{1/m})$$

This equation reveals that the Möbius function, $\mu(n)$, is used in conjunction with $li(n)$ to define Riemann's function, making the connection between Mertens' conjecture and the Riemann hypothesis. Therefore, anyone proving Mertens' conjecture would add a colorful feather to his or her bonnet, proving both conjectures at once. In 1979 H. Cohen and F. Dress computed the values of $M(n)$ for $n$ up to 7.8 billion and still found that Mertens' conjecture held.[12] Finally, in 1984 Andrew Odlyzko and Herman te Riele proved Mertens' conjecture false. They did it by computing the location of the first 2000 nontrivial zeros of the zeta function on the critical line to an accuracy of 100 decimal places. They estimated the first number, $n$, that fails Mertens' conjecture is greater than $10^{30}$, a very large number.[13]

Disproving Mertens' conjecture does not disprove the Riemann hypothesis because the two are not equivalent. The Riemann hypothesis could still be true, even while Mertens' conjecture is false. What this exercise illustrates is that a conjecture which appears to be true for all values calculated to date cannot be assumed to be true on that basis alone. We still don't know the number for which Mertens' conjecture first fails, we simply know that the number is out there somewhere.

## HOW CLOSE?

How close have we come to proving the Riemann hypothesis? G.H. Hardy loved the hypothesis.[14] Certainly he realized its importance to mathematics and recognized its intrinsic beauty, but he may have also enjoyed it because the conjecture appears to have no applications, for useful applications diminished mathematics in his eyes. He would have been dismayed to learn that the Riemann hypothesis has been applied to the science of pyrometry—the study of the internal temperature of furnaces.

Computing the location of individual zeros of $\zeta(s)$ on the critical line is no easy task, for the extended Riemann zeta function is quite involved. J.P. Gram, in 1903, computed the location of the first 15 zeros and found they were all on the critical line. In 1918, R. Backlund increased this to the first 200 zeros—all still on the critical line. Riemann's hypothesis was holding up. As the years progressed, the location of more and more nontrivial zeros were computed, and all lay on the critical line within the critical strip. Of course, we know that the hypothesis can never be proven with this method, however, it can be disproved if only one nontrivial zero is found to be off the line. With modern computers, the calculation of the location of the zeros took a giant leap. By 1983 J. van de Lune and Herman te Riele had found the first 300,000,001 zeros—all still satisfying Riemann's hypothesis. In 1985, they consumed over 1000 hours on a supercomputer to extended this to 1.5 billion zeros, finding no zero off the line.[15]

Is the Riemann hypothesis true or not? We still don't know. By finding segments of the critical strip that are zero-free (excepting the critical line), we can improve the error estimate between $\pi(n)$ and $R(n)$. Under the assumption that the Riemann hypothesis is true, the error decreases to zero much faster, which in turn implies that the primes distribute themselves within the natural number sequence in a much more orderly manner than is presently apparent. Now you know the deepest problem of mathematics. Perhaps to prove or disprove the Riemann hypothesis we will have to wait for another Ramanujan or Hardy, or even Riemann, to come along. However, another possibility exists: What if we can't prove or disprove the Riemann hypothesis? Are there statements within mathematics that we can never prove to be true or false? We just assume that if we can pose a question, then some day we will be able to answer it. Yet, as the forthcoming material will demonstrate, this is not always the case.

Of course, if tomorrow a bright young graduate student did prove the Riemann hypothesis, then its proof would inevitably open up new and deeper questions. This is the way of mathematics.

## CHAPTER FOURTEEN

# INTO THE STRATOSPHERE

*I doubt not but it will be easily granted, that the*

*knowledge we have of mathematical truths is not*

*only certain, but real knowledge; and not the bare*

*empty vision of vain, insignificant chimeras of the*

*brain . . .*[1]

JOHN LOCKE (1632–1704)

*AN ESSAY CONCERNING HUMAN UNDERSTANDING*

## PROVE IT, I DARE YA!

*W*e have discussed many ideas in mathematics, and particularly ideas that have not yet been rigorously proven by mathematicians. For example, we don't know if an infinity of twin primes exist, if either the Goldbach Conjecture or Riemann hypothesis is true. Many of us believe that given enough time and work, all these questions can be answered. Yet, is that the case? Given a statement in mathematics, can we say it is always possible to either prove or disprove it?

A mere 65 years ago, mathematicians believed that any conjecture or question in mathematics could, eventually, be answered true or false. Then, in 1930 a lecture given by a young mathematician was so revolutionary that it shook the very foundation of mathematics until the cornerstones cracked. But first some background.

One of the great gifts of ancient Greek mathematics was Euclid's axiomatic geometry. With a small set of intuitively true axioms, Euclid deduced a complete system of geometry. Could all of mathematics follow this example? As we've shown, during the 19th century Peano produced a foundational system for arithmetic with his five axioms. By the end of the century mathematicians were greatly excited by the prospect that a logical system would soon be found from which all of mathematics could be deduced from some finite set of axioms using the finite rules of pure logic. If mathematicians could show that such a set of axioms was both consistent and complete, then mathematics would, for all practical purposes, be conquered.

Consistency means that it is impossible to deduce a contradiction from the axioms. If we use a set of axioms that produces a contradiction, then we can prove that all statements in the system are true! This won't do because when there is no falsehood and every statement is true, the whole system becomes useless. If we can deduce every possible statement, then what distinguishes truth from falsehood? Therefore, any set of axioms must be consistent to be the foundation stone for mathematics.

The next requirement for any potential set of axioms is completeness. To be complete, a system of axioms must be able to generate a proof for all true statements and disprove all falsehoods which can be formulated under the logical rules adopted. If we have statements that cannot be determined to be either true or false, then our system is incomplete.

The desire to create such a complete, consistent system of mathematics was kindled during the rise of three related philosophies of mathematics. At the turn of the century the University of Vienna hosted a group of prestigious thinkers from various fields, known as the Vienna Circle, whose members developed a philosophical doctrine known as *Logical Positivism*. Part of this philosophy was the desire to unify all of science (especially mathematics) under the language of symbolic logic. We see examples of the Logical Positivist's goal alive today in the attempt by physicists to find one unifying theory for energy and matter. Among logicians and mathematicians, the desire to place all of mathematics under logic was known as *Logicism*, and advocated by such giants as Bertrand Russell and Ludwig Wittgenstein. Closely allied to this philosophy is *formalism*, the belief that mathematics is really a meaningless human game of manipulating symbols with the use of arbitrary but well-defined rules. The aim of the game is to be consistent.[2] Formalism's most famous advocate was David Hilbert (1862–1943), who, you will remember, listed 23 mathematical problems for the 20th century. His second problem was to prove the consistency of arithmetic. If this could be achieved, then we could always be assured that it would be impossible to produce a contradiction from the system.

Bertrand Russell and Alfred North Whitehead made the most successful effort at Formalism, attempting to prove the consistency of arithmetic using Peano's axioms. They almost succeeded. They were successful in eliminating vagueness from the rules of logic, reducing them to a small set of *Transformation Rules* that allows one to move securely from a set of premises (axioms and theorems) to deduce new theorems. In their effort they produced a monumental three volume book, *Principia Mathematica*, 1910–1913.

We can show a system is consistent when the objects within that system are finite in number. However, arithmetic contains an infinite number of numbers, and when we try to clearly describe that system, concealed contradictions, or antinomies, pop out. Bertrand Russell discovered such a contradiction in set theory, the foundation to mathematics. We can illustrate Russell's contradiction with the following Barber Paradox.

Suppose there is a town with only one barber. This barber shaves all the men and only the men who do not shave themselves. Now, who shaves the barber? If he does not shave himself, then he falls into the group of men he is supposed to shave. If he does shave himself, then he shouldn't shave himself—a paradox. Because of the discovery of antinomies within the logical foundation of mathematics, many feared that such contradictions might infect other branches of mathematics. Hence, Russell and others went to work to build a kind of axiomatic theory that would avoid the problem.

Not much happened for the next decade or so as mathematicians and logicians tried to dance around the antinomies while still holding firm to the belief that the proof of both consistency and completeness were on the horizon. Suddenly enters Gödel.

## ENTERS GÖDEL

Kurt Gödel was unquestionably the greatest logician of the century. He may also have been one of our greatest philosophers.[3]

RUDY RUCKER

[Gödel's Proof] is generally regarded as the most brilliant, most difficult, and most stunning sequence of reasoning in modern logic.[4]

JAMES R. NEWMAN

This proof, by the Austrian Kurt Gödel in 1931, is one of the most remarkable and devastating discoveries in the whole of mathematics.[5]

J.M. DUBBEY

Kurt Gödel was born on April 28, 1906 in Brünn, Czechoslovakia, which was then part of the Austria-Hungary Empire. His family was German, and his father was a manager for one of the city's textile mills.[6] At 6, Gödel contracted rheumatic fever, which gave him an obsessive fear about his health in later life. In 1923 he entered the University of Vienna where he came into contact with the ideas of Logical Positivism and Logicism as advanced by the Vienna Circle. In 1930 he earned his doctorate in mathematics.

On September 7, 1930, Kurt Gödel, then only 24, presented a 22-minute talk at a conference at Königsberg, Russia, entitled "On Formally Undecidable Propositions of Principia Mathematica and Related Systems."[7] At the time no one seemed to take much notice of the young mathematician's talk, except the American mathematician John von Neumann, cofounder of game theory and the father of the modern computer (he devised computer programs that could alter themselves while running on a computer). After the conference, von Neumann approached Gödel and asked for more details on his logical argument. In 1931, Gödel's work was published, and suddenly, the cat was out of the bag. What in the world had Gödel proven?[8]

## THE PROOF

Be warned this next section may require some quiet, reflective thought, for Gödel's proof is not easy. The original paper contained 46 preliminary definitions and several lemmas before the final work could be approached.[9] Yet, we do not shirk from our quest.

Russell and Whitehead had used formal logic as a foundation for set theory, which in turn was the formal basis for arithmetic. The two questions which had not yet been answered were: first, are the axioms consistent, and second, were the axioms complete? The young Czech from Vienna proved a limited case of inconsistency. He proved that if formal set theory is complete, then it is ω-inconsistent. In 1936, J. Barkley Rosser expanded Gödel's proof to give us the following general theorem.

> *Gödel-Rosser Theorem:* If formal set theory is consistent then theorems exist that can be neither proved nor disproved.

Certainly, thought most mathematicians, what Gödel has proven is that theorems exist which cannot *now* be proven, but, given enough time and effort, they will fall to our assault. But, no, this was not what Gödel had done. What he had actually demonstrated was that it was *logically impossible to ever prove* the theorems in question within formal set theory. That is, they were theoretically unprovable if we limit ourselves to the axioms of set theory.

If set theory is not consistent, then it is worthless as a basis to anything, for one can prove everything and anything using an inconsistent system—inconsistent systems cannot distinguish between truth and falsehood. On the other hand, if set theory was consistent, then true statements existed within set theory which, according to Gödel, could never, ever be proven to be true. That is, if set theory is consistent, then it cannot be complete.

How had he done it? Gödel's proof is an intricate dance of logic that actually constructs a theorem that is true, but not provable within the system. The system representing the theoretical basis for arithmetic can be divided into several parts. First there are the laws of logic, which Whitehead and Russell had helped to formalize. Next there are the axioms of the system used to deduce all the theorems that specify the arithmetic laws regarding numbers. These axioms and theorems are written in a very specific and limited symbolism to avoid ambiguity. Remember when we looked at Peano's axioms, the second axiom read: Every number has a successor which is also a number. We can represent this axiom in symbolic form as: $(\exists x)(x = sy)$. In this symbolic statement, the backward capital $e$ stands for "there exists," and the small $s$ stands for "the successor of," while the other two letters, $x$ and $y$, are individual variables. Hence, we interpret the symbolism as "There exists an $x$ (a number) such that $x$ is the successor of $y$, where $y$ is any number." Therefore, given any number, $y$, we are guaranteed a successor of $y$, which is just what we need when we talk about the natural numbers. If someone gives us a natural number, no matter how large, we always know that there exists another number that is one unit larger (the successor).

The symbolic statement [e.g., $(\exists x)(x = sy)$] is called the "system calculus" while the English interpretation of the symbolic statement is called the meta-mathematical statement. The meta-mathematical statement gives meaning or content to the symbolic statement. When we perform mathematics by using the system calculus, we manipulate the symbolic statements by means of the logical rules or Transformation Rules. What Gödel did was to construct a symbolic statement whose meta-mathematical statement was true. At the same time, the very construction of the symbolic statement guaranteed that it could never be proven within the system.

When we prove a statement in symbolic logic, we list the axioms and those theorems that have already been proven from the axioms that we need to deduce the conclusion. The axioms and theorems are called premises. Hence, we have a list of premises followed by the conclusion or:

Premise 1
Premise 2
Premise 3
⋮
Premise $n$
Conclusion

All the $n$ premises logically lead to the conclusion, and in fact, the conclusion is implied by the premises.

In order to construct his true, nondeducible statement, Gödel devised a method to assign a unique number to every symbol within the system, and a unique number to every statement and every set of statements that represented a deduction of a theorem. These numbers are now referred to as Gödel numbers. To accomplish this he used a scheme similar to the following:

| Logical Symbol | Gödel Number | Meaning |
|:---:|:---:|:---:|
| ~ | 1 | negation or "not" |
| v | 2 | "or" |
| ⊃ | 3 | "if . . . then" |
| ∃ | 4 | "there exists" |
| = | 5 | "equals" |
| 0 | 6 | zero |
| s | 7 | immediate successor |
| ( | 8 | left parenthesis |
| ) | 9 | right parenthesis |
| , | 10 | comma |

In addition to the above, the system uses numerical variables ($x$, $y$, $z$) which can take the values of specific numbers, sentential variables ($p$, $q$, $r$) which stand for other symbolic statements, and predicate variables ($P$, $Q$, $R$) which stand for predicates, all of which are given unique values of numbers greater than 10. Each numerical variable is assigned the value of a prime greater than 10, each sentential variable is assigned the square of a prime greater then 10, while each predicate variable is given the cube of a prime greater than 10. Under this system we might assign $x$ the value of 11, $p$ the value of $11^2$, and $P$ the value of $11^3$.

Now we're ready to assign the Gödel numbers to the statement $(\exists x)(x = sy)$. Reading left to right, the Gödel numbers for the ten symbols in this statement are: 8, 4, 13, 9, 8, 13, 5, 7, 16, and 9. In order to create a unique integer that is associated with only our statement, and none other, we take the first ten successive prime numbers and raise each to the corresponding Gödel number, and then multiply the whole mess together. Hence, we get:

Gödel number for $(\exists x)(x = sy) = 2^8 3^4 5^{13} 7^9 11^8 13^{13} 17^5 19^7 23^{16} 29^9$

Now, admittedly, this is one gigantic number. However, it is uniquely mapped onto our symbolic statement. If someone gave us this number all multiplied out, we could factor it into its unique primes, order the primes from smallest to largest, and get the above number. We could then easily assign the appropriate symbols and retrieve our axiom about numbers. Therefore, Gödel's procedure uniquely assigns an integer to each logical statement within the system.

We should realize that every integer is not a Gödel number; the Gödel numbers represent only a subset of the integers. For example, the number 2400 factors into $2^5 \cdot 3^1 \cdot 5^2$ which has the corresponding symbolism of $= v \sim$, which in turn can be read as "equals not or." But neither the symbolic statement of $= v \sim$, nor the meta-mathematical interpretation of "equals not or" have any meaning because they are not well-formed according to the Transformation Rules. Another example is the number 70 which factors into $2^1 \cdot 5^1 \cdot 7^1$. This number does not contain the prime 3, so the number corresponding to the second symbol of a potential symbolic statement is missing. Hence, 70 cannot be a Gödel number. Only those numbers that factor into consecutive primes, beginning with 2, raised to a power, resulting in well-formed statements have Gödel numbers.

The next step is to list the statements that are premises leading to a specific conclusion. This premises–conclusion list of statements is a logical argument. Each of the premises and the conclusion have a unique Gödel number and, in the same fashion, we construct a new Gödel number which corresponds to the entire argument. Suppose we have an argument of the form: premise 1, premise 2, and conclusion. Let the Gödel numbers for these three statements be $J$, $K$, and $M$. Now we select the first three prime numbers (2, 3, and 5) and form the following number:

$$\text{Gödel number for argument} = 2^J 3^K 5^M$$

In addition we can also compute the Gödel number for just the two premises by themselves if we want, which would be:

$$\text{Gödel number for premises}, P_G = 2^J 3^K$$

Using this procedure, every true statement and argument within the logical system gets its unique integer. Yet, each symbolic statement also has its corresponding meta-mathematical statement. This means that our meta-mathematical statements about the system can be mapped into the system, itself. We now have three related parts to our system: symbolic statements, corresponding meta-mathematical statements, and corresponding Gödel numbers.

When we assign our Gödel numbers, we discover something quite wonderful: There is a definite arithmetic relation between the Gödel numbers for the premises, $P_G$, and the Gödel number for the conclusion, $M$. For example, $5 > 2$ is a true arithmetic relation between 5 and 2. In a similar fashion there exists a true arithmetic relation between the Gödel numbers of the premises and conclusion of every valid argument. These relationships are rather complex, and part of Gödel's genius is that he could deal with them. While we are not going to compute all the Gödel numbers in his proof and detail their relationships, we can offer an example. One of the logical axioms used is $(p \, v \, p) \supset p$. This simply says that if either the statement $p$ or $p$ is true, then $p$ is true. When we compute Gödel's number for this statement we get:

$$\text{Gödel number } (p \, v \, p) \supset p = 2^8 \cdot 3^{11^2} \cdot 5^2 \cdot 7^{11^2} \cdot 11^9 \cdot 13^3 \cdot 17^{11^2}$$

Now, this statement, $(p \vee p) \supset p$, can be viewed as an argument
where $(p \vee p)$ is the premise and $p$ is the conclusion. If we compute
the Gödel number for just $(p \vee p)$ we get:

$$2^8 . 3^{11^2} . 5^2 . 7^{11^2} . 11^9$$

which evenly divides the larger Gödel number represented by the
entire statement, and hence is a factor of the entire statement. For
a number to be a factor of a larger number is an arithmetic property.
This demonstrates that Gödel's procedure for assigning Gödel
numbers to logical deductions within the system translates to
corresponding arithmetic relations between the Gödel numbers. In
fact, if the symbolic statement is true (deducible from within the
system) then the corresponding arithmetic relationship between
the Gödel numbers representing the symbolic statement and its
premises will also be true.

We are now ready for the next step. The Gödel numbers are not
only an arithmetic reflection of the symbolic statements, but they
can also be a reflection for the meta-mathematical statements corresponding to the symbolic statements. We can designate the arithmetic relation between Gödel numbers representing a set of
premises and a conclusion as $\mathrm{Dem}(x,y)$, where $x$ is the Gödel
number for the premises and $y$ is the Gödel number for the conclusion. Keep in mind that $\mathrm{Dem}(x,y)$ is not a symbolic statement, but
is an arithmetic relationship (which is quite complex) existing
between the Gödel number for all the premises, $x$, and the Gödel
number for the conclusion, $y$. We can also form the denial of
$\mathrm{Dem}(x,y)$ which is shown as $\sim\mathrm{Dem}(x,y)$ which means that there is
no arithmetic relationship between $x$ and $y$ where $x$ represents the
premises and $y$ represents the conclusion of an argument. However,
if there is no arithmetic relationship between the Gödel numbers
of the premises and conclusion, this is equivalent to saying that
there exists no set of premises within the system that have $y$ as a
conclusion. This takes a little thought to get clear. Every valid
argument has a set of premises and a resulting conclusion. The
corresponding Gödel numbers for the premises and conclusion, $x$
and $y$, have a true arithmetic relationship. If no such arithmetic

relationship exists between $x$ and $y$, then the corresponding premises do not lead to the conclusion. We have one more definition to make before we can show the final step. If we have a symbolic statement which contains a numerical variable, we can substitute some specific number for the variable. For example, the Peano axiom $(\exists x)(x = sy)$ says that there exists some $x$ that is the successor for $y$. We may substitute the specific number 5 for $y$. When we make this substitution we get $(\exists x)(x = s5)$, which says that there exists a successor to the number 5. We know this is true, because the successor is, in fact, 6. Whenever we make such a substitution we generate a new Gödel number because the Gödel number for $(\exists x)(x = s5)$ is different than the Gödel number for $(\exists x)(x = sy)$. We did not assign a Gödel number to 5. However, we can determine one by remembering that all whole numbers can be defined as successors to other numbers. Hence, we have 1 is $s0$, 2 is $s1$, 3 is $s2$, and so forth, where $s$ stands for "successor of." Once we do this, every whole number is represented by a series of $s$'s followed by 0. For example, a Gödel number for 5 can be defined by $sssss0$, since we already have the Gödel numbers for $s$ and 0.

When we generate a new Gödel number by making a substitution for one of the numerical variables, we show this number as $\text{sub}(x, y, z)$ where $x$ is the Gödel number of the original statement, $y$ is the term we're substituting for, and $z$ is what we are substituting. Hence, in our previous example we have $\text{sub}((\exists x)(x = sy), y, 5)$. We could have used 13 for $y$ since 13 is the Gödel number for $y$. We interpret this particular sub statement as the Gödel number that results when we substitute the number 5 for the $y$ into the statement $(\exists x)(x = sy)$. Let's take this idea of substitution one step further. Say the Gödel number for $(\exists x)(x = sy)$ happens to be $m$. Now we can make our substitution using $m$ or $\text{sub}(m,y,m)$. This says, find the new Gödel number when we substitute $m$ for $y$ in the statement that has Gödel number $m$. You see, $m$ uniquely identifies the statement $(\exists x)(x = sy)$, and we are going to substitute $m$ back into the very statement that $m$ identifies and we are substituting $m$ for the variable $y$. We then compute the new Gödel number for the statement resulting from the substitution. However, we're not going to actually compute this number, but only show it as the number sub($m,y,m$). We have introduced so many new terms and ideas your head is probably swimming. Remember that we now have four different entities in our game:

1) Symbolic statements
2) Meta-mathematical statements
3) Gödel numbers
4) Arithmetic relations between Gödel numbers

We can now form Gödel's true but unprovable statement. We begin by first stating the following rather complex statement:

$$(x) \sim \text{Dem}(x, \text{sub}(y,13,y))$$

This remarkable statement says the following: for every set of premises, represented by $(x)$, there does not exist a true arithmetic relationship between $x$ and the number sub($y,13,y$), demonstrating that sub($y,13,y$) is the conclusion of $x$. This is the same as saying that the conclusion, sub($y,13,y$), cannot be deduced in the system, because no set of premises will lead to it as a conclusion. Hence, there is no proof of sub($y,13,y$) within the system. Now, does $y$ exist, and is it true? If we can show that there is a $y$ which is true and cannot be deduced from any set of premises, we will have shown that our system is incomplete, i.e., there exists a true statement which cannot ever be given a proof.

What is $y$? Now we see Gödel's true genius. Let $n$ be the Gödel number for the above statement, $(x) \sim \text{Dem}(x, \text{sub}(y,13,y))$. Now we form Gödel's brilliant conclusion (a little drum roll please).

$$(x) \sim \text{Dem}(x, \text{sub}(n,13,n))$$

We will call this the $G$ statement. Now we should take some time just to feast our eyes on the marvelous construction. What is the Gödel number of the $G$ statement? It is simply sub($n,13,n$). Hence, the statement says that the formula, $(x) \sim \text{Dem}(x, \text{sub}(n,13,n))$, is not demonstrable within the system; it asserts it own undemonstrability. To see this we must remember just what number sub($n$,13,$n$) is: it is the Gödel number for the above statement. Hence, $G$ is not the conclusion to any set of premises. But is $G$ true? We can prove it is true through the logic of meta-mathematics. Suppose $G$ is false. If $G$ is false, then there exists a set of premises within the system (that have a Gödel number of $x$) with $G$ as its conclusion (since the Gödel number of $G$ is sub($n$,13,$n$)). But that means that $G$ is true! Therefore, if we assume $G$ is false, then we must conclude that it is true—a contradiction. If a system can prove a contradiction, then it is an inconsistent system. Therefore, if our system is consistent, we cannot conclude that both $G$ is true and false. Hence, if our system is consistent, then $G$ cannot be false. Since $G$ cannot be false, it must be true. But, if it is true, then it says that it cannot be deduced within the system. What brilliance!

## WHAT DOES IT ALL MEAN?

Gödel proved that formal arithmetic was incomplete if it was consistent. There will always be unprovable statements which can be generated that are beyond formal proof. Have we encountered some already? Is the Riemann hypothesis unprovable? What about the Goldbach Conjecture?

With his proof, Gödel showed that Hilbert's program for formalizing all of mathematics was impossible. This did not go well with Hilbert. Some say he blew his top when he heard of Gödel's proof.[10] Even in his later years, Hilbert refused to believe what Gödel had accomplished, holding on to the hope that the consistency and completeness of set theory and arithmetic could be demonstrated.[11] Despite Hilbert's reluctance to abandon the Formalist program, Gödel's proof is now accepted by the vast majority of mathematicians.

In 1933 Gödel dropped a second bomb on theoretical mathematics. He proved that no procedure existed for establishing the consistency of any axiomatic system large enough to generate arithmetic. Therefore, not only was Russell's system, based on the Peano axioms and the Transformation Rules, beyond a proof of consistency, but *any* axiomatic system that was sophisticated enough to handle arithmetic could not be proved consistent. The two great flowers of formal mathematics had now wilted beyond hope. There would be no consistency nor completeness for mathematics, if we insisted on using finite logic as defined by the Formalists' program for mathematical proof.

In 1938, Gödel left when Europe was about to be torn apart by war, and permanently joined the Institute for Advanced Study at Princeton. This institute had been established around Albert Einstein a few years before with financial help from the department-store baron, Louis Bamberger.[12] Gödel was a private, shy man, but became close friends with Einstein. During the 1940s he published his one book, *The Consistency of the Continuum Hypothesis*, which attempted to settle another of Hilbert's 23 questions.

As the years progressed, Gödel seemed to become more of a recluse, and numerous stories surfaced regarding his strangeness. For example, he is reported to have gladly accepted invitations to meet with others, but generally failed to make the appointment. When asked why he did this, his response was characteristically logical: He claimed that it was the only way that guaranteed he would *not* have to meet with the other individuals.

Gödel's United States citizenship required two sponsors. Albert Einstein and Oskar Morgenstern (the cofounder of game theory) stepped forward. However, at the citizenship interview, Gödel, in the presence of the government interviewer and his two illustrious sponsors, launched into an agitated lecture on how he had discovered a logical flaw in the United States Constitution, which might allow the United States to be taken over by a dictator. Finally, his two friends were able to calm him down, and his citizenship was granted.

Such idiosyncrasies may sound lovable, yet there is a touch of sadness about his last years. He gradually became obsessed with fears regarding his own health, and may have actually starved himself to death fearing someone was trying to poison him.[13] In his later years, because of his fame, many mathematicians wrote to him, and even though he generally drafted precise responses, they were seldom actually mailed. After his death, his private papers demonstrated a great store of work on the philosophy of mathematics, as well as general philosophical issues, all written in a strange Germanic shorthand. He was not only a great thinker about logic and formal mathematics, he had a powerful mind that could shine its light on most subjects.[14]

## SO PHILOSOPHICAL

Because of the popularity of Logical Positivism in philosophy and Formalism within mathematics at the turn of the century, the prevailing view was that the objects of mathematics, including numbers, were objects whose entire existence were determined by individual human minds. They only existed in human thought, and would entirely cease to exist should all minds disappear. The objects of mathematics were like conventions of a made-up game, depending entirely upon individual thinking minds. Yet, there was an older philosophical idea regarding mathematics alive at that time which can be traced back to Plato and Pythagoras: idealism, or Platonism, which maintains that mathematical objects existed before humans, are not dependent on humans, and will exist long after we are gone. A Platonist contemporary of Gödel was G.H. Hardy.

Deciding between mathematical Platonism and Formalism is difficult. Gödel, however, stood firmly in the camp of Platonism, so much so that modern mathematical Platonism only began to gather a significant number of disciples after Gödel's influence in the 1930s. Along with Einstein, he was somewhat of a mystic, believing that we somehow perceive mathematical objects with our reason, just as our senses allow us to perceive sensory phenomena. Gödel used to say that he practiced "objective" mathematics, to emphasize that he perceived or discovered mathematical relations, rather than inventing them.[15] While many modern mathematicians talk as if they are Formalists, declaring that mathematics is only in human minds, they, in fact, act more like Platonists, frequently surprised by their mathematical "discoveries." A current Platonist is Roger Penrose, who asks us to consider the beautiful fractal images discovered buried within certain mathematical (and physical) processes.[16] Could such images really be nothing but constructs of the inventive mind?

We can now recognize a unifying idea in both Russell's paradox, as illustrated by the barber paradox, and Gödel's proof of the incompleteness of mathematics. Both contain statements that refer to themselves. In the barber paradox, the paradox is generated when the barber must consider cutting his own hair. Gödel's proof is built on the notion that a symbolic statement can talk about itself. Douglas Hofstadter, a mathematician, philosopher, psychologist, and all-around interesting person, has extensively studied self reference.[17] Some of the examples he has created or collected from others are quite amusing, while others suggest a strange deepness within self reference. The first recorded self reference comes to us from ancient Greece. Epimenides (fl. sixth and fifth centuries B.C.) is reported to have said, "The Cretans are always liars, evil beasts, lazy stomachs."[18] However, Epimenides, himself, was Cretan, and therefore his statement about Cretans always being liars must also apply to him. This amusing paradox has changed through the ages to become: "This sentence is false." If it is false, then it must be true, and if true, then it is false.

A small sampling of Hofstadter's collection illustrates how intriguing self reference is.

This sentence no verb.

Well, how about that—this sentence is about me!

I am the thought you are now thinking.

The reader of this sentence exists only while reading me.

When you are not looking at it, this sentence is in Spanish.[19]

However, such sentences are not for play only, for they can demonstrate that self reference can suggest profound ideas. We have already seen this in both the Russell paradox and Gödel's proof. One last self reference will do, again from the fine Hofstadter collection: "What would this sentence be like if π were 3?"[20] Indeed, we might wonder if this sentence has any meaning at all, yet we all read and understand its meaning. Can such a question be asked in all seriousness? Why would we suspect that if $\pi$ were 3, instead of the strange transcendental number that it is, that anything else in the world would be different? If we assume that a difference in the value of $\pi$ makes a difference in reality, then are we not succumbing to the Platonist platform?

## WHERE DO WE STAND?

Surprisingly, a proof of the consistency of arithmetic has been achieved. In 1936 Gerhard Gentzen proved the consistency using logical rules which included propositions involving infinite premises.[21] This goes considerably beyond the original idea that the Transformation Rules could be finite, and opens up questions of whether these expanded rules of logic are, themselves, consistent.

Gödel's work has shown that the arithmetic invented (discovered?) by the human species is far too complex and powerful to be tamed by a finite set of axioms and logical rules. Yet, this is not the only potential limitation to our understanding of mathematics. We now know that there exist true propositions which we can never formally prove. What about propositions whose proofs require arguments beyond our capabilities? What about propositions whose proofs require millions of pages? Or a million, million pages? Are there proofs that are possible, but beyond us? Perhaps. But remember that the poor chimp cannot understand even a single mathematical proof, while we teach proofs to our children. Maybe the human species will evolve into another, higher species, and they will grasp the answer to all these questions as simply as we add 1 and 1 to get 2. It makes one wonder.

## THE LAST STRAW

This has been such a serious chapter. We must remember that our quest of mathematics is one of joy and fun. In harmony with this philosophy we offer once again the sage words of Thomas Paine.

> The mere man of pleasure is miserable in old age, and the mere drudge in business is but little better, whereas, natural philosophy, mathematical and mechanical science, are a continual source

of tranquil pleasure, and in spite of the gloomy dogmas of priests and of superstition, the study of these things is the true theology; it teaches man to know and to admire the Creator, for the principles of science are in the creation, and are unchangeable and of divine origin.[22]

Let's end with a riddle. Below is a simple number sequence. Think about it, ponder on it, and devour it. When you have discovered the key, turn to the last page and confirm your answer to the wonderful Conway sequence (invented by the English mathematician, John Horton Conway, of Cambridge University).

$$1$$
$$11$$
$$21$$
$$1211$$
$$111221$$
$$?$$

# END NOTES

## INTRODUCTION

[1] E.T. Bell, *Men of Mathematics* (New York: Simon & Schuster, 1937), p. 405.

[2] Stephen Hawking, ed., *A Brief History of Time: A Reader's Companion* (New York: Bantam Books, 1992), p. vii.

[3] Internet: http://www.groups.dcs.st-and.ac.uk:80/~history/

[4] Internet: primes@math.utm.edu

[5] Internet: wuarchive.wustl.edu/doc/misc/pi

## CHAPTER 1

[1] Mark Twain, *The Adventures of Huckleberry Finn* (Franklin Center, PA: The Franklin Library, 1983), p. 19.

[2] Donald R. Griffin, *Animal Thinking* (Cambridge, MA: Harvard University Press, 1984); Guy Woodruff and David Pemack, "Primate Mathematical Concepts in the Chimpanzee: Proportionality and Numerosity," *Nature*, Vol. 293, October 15, 1981, 568.

[3] Recent evidence suggests that *Homo erectus* may be much older—as much as 2.5 million years old.

[4] Denise Schmandt-Besserat, *Before Writing* (Austin, TX: University of Texas Press, 1992).

[5] Jacques Soustelle, *Mexico* (New York: World Publishing Company, 1967), p. 125.

[6] Peano's axioms actually contain the primitive term "zero" rather than "1." For this illustration I have begun the number sequence with 1 rather than zero.

[7] Kathleen Freeman, *Ancilla to the Pre-Socratic Philosophers* (Cambridge, MA: Harvard University Press, 1966), p. 75.

[8] Sir Thomas Heath, *A History of Greek Mathematics* (London: Oxford University Press, 1921), p. 75.

[9]Aristotle, *The Basic Works of Aristotle*, trans. J. Annas (Richard McKoen, ed.) (New York: Random House, 1941); *The Metaphysics*, 986a, lines 15–18, Oxford University Press.

[10]James R. Newman, "The Rhind Papyrus," in *The World of Mathematics, Vol. 1*, ed. James R. Newman (New York: Simon and Schuster, 1956), p. 174.

[11]Heath, *A History of Greek Mathematics*, p. 76.

[12]Internet: sci.math, Alex Lopez-Ortiz, University of Waterloo, alopez-o@maytag.UWaterloo.ca, 6/16/94.

[13]Ibid.

[14]Philip J. Davis, *The Lore of Large Numbers* (New York: Random House, 1961), p. 23.

[15]Internet: sci.math, Lee Rudolph, Department of Mathematics, Clark University, rudolph@cis.umassd.edu, 6/27/94.

## CHAPTER 2

[1]H.G. Wells, *War of the Worlds* (CD: DeskTop BookShop) (Indianapolis: WeMake CDs, Inc., 1994).

[2]Leviticus 4:6, *The Holy Bible, King James Version* (New York: The World Publishing Company), p. 78.

[3]Joshua 6:4, *The Holy Bible, King James Version* (New York: The World Publishing Company), p. 165.

[4]González-Wippler, *The Complete Book of Spells, Ceremonies, and Magic* (New York: Crown Publishers, Inc., 1978), p. 333.

[5]Ibid, p. 344.

[6]David Eugene Smith, *History of Mathematics* (New York: Dover Publications, Inc., 1951), p. 29.

[7]Rosemary Ellen Guiley, *The Encyclopedia of Witches and Witchcraft* (New York: Facts On File, Inc., 1989), p. 135.

[8]Revelation 13:18, *The Holy Bible, King James Version* (New York: The World Publishing Company), p. 189.

## CHAPTER 3

[1]Alfred North Whitehead, *An Introduction to Mathematics* (New York: Oxford University Press, 1958), p. 144.

[2]Konrad Knopp, *Theory and Application of Infinite Series* (New York: Dover Publications, 1990), p. 67.

[3]Florian Cajori, *A History of Elementary Mathematics* (New York: Macmillan & Company, 1930), p. 100; David Eugene Smith, *History of Mathematics* (New York: Dover Publications, 1951), p. 266.

[4]Florian Cajori, *A History of Elementary Mathematics*, p. 9.

[5]Gay Robins and Charles Shute, *The Rhind Mathematical Papyrus* (New York: Dover Publications, 1987), p. 42.

## CHAPTER 4

[1]Francis Bacon, *Essays* (CD: DeskTop BookShop) (Indianapolis: WeMake CDs, Inc., 1994).

[2]Lucas Bunt, Phillip Jones, and Jack Bedient, *The Historical Roots of Elementary Mathematics* (New York: Dover Publications, 1976), p. 86.

[3]Florian Cajori, *A History of Elementary Mathematics* (New York: Macmillan & Company, 1930), p. 44.

[4]David Eugene Smith, *History of Mathematics* (New York: Dover Publications, 1951), p. 143.

[5]א (aleph) is the first letter of the Hebrew alphabet and is commonly used to represent infinite sets.

## CHAPTER 5

[1]Ralph Waldo Emerson, *Representative Man* (CD: DeskTop BookShop) (Indianapolis: WeMake CDs, Inc., 1994).

[2]The first to prove this remarkable result was actually the Englishman, Richard Suiseth (fl. ca. 1350), who gave a long verbal proof of an equivalent expression.

[3]Others were also working along the same lines at this time. In fact, Jobst Bürgi (1552–1632) of Switzerland may have developed a similar system as early as 1588, but did not publish his ideas until 1620, after the work of Napier had already appeared.

## CHAPTER 6

[1]Thomas Paine, *Address to the People of England* (CD: DeskTop BookShop) (Indianapolis: WeMake CDs, Inc., 1994).

[2]Carl B. Boyer, *A History of Mathematics* (New York: John Wiley and Sons, 1968), p. 55.

[3]H.E. Huntley, *The Divine Proportion* (New York: Dover Publications, 1970), p. 62.

[4]David Wells, *The Penguin Dictionary of Curious and Interesting Numbers* (New York: Penguin Books, 1986), p. 37.

[5]Boyer, *A History of Mathematics*, p. 281.

[6]*The Fibonacci Quarterly*, Fibonacci Association c/o South Dakota State University Computer Science Department, Box 2201, Brookings, SD 57007-1596.

[7]H.E. Huntley, *The Divine Proportion*, p. 160.

[8]George Cheverghese Joseph, *The Crest of the Peacock* (London: Penguin Books, 1991), p. 197.

[9]Jean L. McKechnie, ed., *Webster's New Twentieth Century Dictionary of English Language* (New York: Simon and Schuster, 1979), p. 1834.

## CHAPTER 7

[1]H.G. Wells, *The Time Machine* (CD: DeskTop BookShop) (Indianapolis: WeMake CDs, Inc., 1994).

[2]Paulo Ribenboim, *The Little Book of Big Primes* (New York: Springer-Verlag, 1991), p. 142.

[3]For a further discussion of this equation see Paulo Ribenboim, *The Book of Prime Number Records*, second edition (New York: Springer-Verlag, 1989), p. 190.

## CHAPTER 8

[1]John Locke, *An Essay Concerning Human Understanding* (CD: DeskTop BookShop) (Indianapolis: WeMake CDs, Inc., 1994).

[2]Paulo Ribenboim, *The Book of Prime Number Records*, second edition (New York: Springer-Verlag, 1989), p. 130.

[3]Ibid, p. 136.

[4]Ibid, p. 144.

[5]Since Samuel Yates' death, the list of Titanic primes has been kept by Professor Chris Caldwell who is kind enough to make the list available to interested parties. If you would like a list of the largest primes send $4 to:

Professor Chris Caldwell

Department of Mathematics

University of Tennessee at Martin

Martin, TN 38238

If you have access to the Internet, Professor Caldwell maintains a home page on the World Wide Web containing the largest primes: http://www.utm.edu:80/research/primes/largest.html. The records for largest primes of various types listed here all come from Caldwell's Internet list.

[6]R1031 was discovered by Williams and Dubner in 1986.

[7]Ribenboim, *The Book of Prime Number Records*, p. 286.

## CHAPTER 9

[1]Herodotus, *The History of Herodotus* (CD: DeskTop BookShop) (Indianapolis: WeMake CDs, Inc., 1994).

[2]D. James Bidzos and Burt S. Kaliski Jr., "An Overview of Cryptography," *LAN TIMES*, February 1990.

[3]Whitfield Diffie and Martin E. Hellman, "Privacy and Authentication: An Introduction to Cryptography," *Proceedings of the IEEE*, Vol. 67, No. 3, March 1979, 397–427.

[4]Ronald L. Rivest, Adi Shamir, and Leonard Adleman, "A Method for Obtaining Digital Signatures and Public-Key Cryptosystems," *Communications of the ACM*, Vol. 21, No. 2, February 1978, 120–126.

[5]William Booth, "To Break the Unbreakable Number," *The Washington Post*, June 25, 1990, A3.

[6]Paulo Ribenboim, *The Book of Prime Number Records*, second edition (New York: Springer-Verlag, 1989), p. 478.

[7]Barry A. Cipra, "PCs Factor a 'Most Wanted' Number," *Science*, Vol. 242, December 23, 1988, 1634.

[8]Private phone conversation with Kurt Stammberger, Technology Marketing Manager, RSA Data Security, Inc., June 22, 1995.

[9]I must give special thanks to both Professor Ronald Rivest of MIT and Kurt R. Stammberger, Sales and Marketing Manager for RSA Data Security, Inc., for all the information they provided to me on the RSA cryptosystem and the RSA factorization contest.

[10]RSA Laboratories, *CryptoBytes*, Vol. 1, No. 1, Spring 1995, 1.

[11]Want to try your hand at additional numbers in the challenge? For more information on the contest and a copy of the numbers write to:
RSA Challenge Administrator
100 Marine Parkway, Suite 500
Redwood City, CA 94065
(415) 595-8782
or send e-mail to: challenge-info@rsa.com, or browse the World Wide Web page, http://www.rsa.com.

[12]RSA Laboratories, "Answers to Frequently Asked Questions," revision 2.0, October 5, 1993.

## CHAPTER 10

[1]Henry David Thoreau, *Walden* (CD: DeskTop BookShop) (Indianapolis: WeMake CDs, Inc., 1994).

[2]Robert Kanigel, *The Man Who Knew Infinity* (New York: Charles Scribner's Sons, 1991), p. 11.

[3]Ibid, p. 71.

[4]Ibid, p. 86.

[5]Bruce C. Berndt, *Ramanujan's Notebooks*, Vol. I (New York: Springer Verlag, 1985), p. 2.

[6]Ibid, p. 4.

[7]Robert Kanigel, *The Man Who Knew Infinity*, p. 203.

[8]James R. Newman, ed., *The World of Mathematics*, Vol. 1 (New York: Simon and Schuster, 1956), pp. 371–372.

[9]G. H. Hardy, *A Course of Pure Mathematics* (London: Cambridge University Press, 1963); G.H. Hardy and E.M. Wright, *An Introduction to the Theory of Numbers* (New York: Oxford University Press, 1979).

[10]James R. Newman, ed., *The World of Mathematics*, p. 2029.

[11]Robert Kanigel, *The Man Who Knew Infinity*, p. 347.

[12]James R. Newman, ed., *The World of Mathematics*, p. 2038.

[13]Ibid, pp. 2027–2029.

[14]Jerry P. King, *The Art of Mathematics* (New York: Plenum Press, 1992), p. 29.

## CHAPTER 11

[1]Plato, *The Dialogues of Plato*, trans. B. Jowett (New York: Random House, 1937), Timaeus, p. 66.

[2]Euclid, *Elements, Book V* (New York: Dover Publications, 1956), p. 139.

[3]The majority of the following Ramanujan equations have been taken from Bruce C. Berndt, *Ramanujan's Notebooks, Vol. I & II* (New York: Springer Verlag, 1985).

[4]Robert Kanigel, *The Man Who Knew Infinity*, p. 247.

[5]David Wells, *Curious and Interesting Numbers* (London: Penguin Books, 1986), p. 100.

[6]Konrad Knopp, *Theory and Application of Infinite Series* (New York: Dover Publications, 1990), p. 548.

[7]Ivars Peterson, *Islands of Truth: A Mathematical Mystery Cruise* (New York: W.H. Freeman and Company, 1990), p. 177.

## CHAPTER 12

[1]David Hume, *An Enquiry Concerning Human Understanding* (CD: DeskTop BookShop) (Indianapolis: WeMake CDs, Inc., 1994).

[2]Internet: sci.math, Kent D. Boklan, July 21, 1994.

[3]Paulo Ribenboim, *The Book of Prime Number Records*, p. 230.

[4]Internet: sci.math, Boklan, July 21, 1994.

[5]Paulo Ribenboim, *The Book of Prime Number Records*, p. 230.

[6]Internet: sci.math, Chris Thompson, August 5, 1994.

[7]Henry F. Fliegel and Douglas S. Robertson, "Goldbach's Comet: The Numbers Related to Goldbach's Conjecture," *Journal of Recreational Mathematics*, Vol. 21, No. 1, 1989, 1–7.

[8]G.H. Hardy and E.M. Wright, *An Introduction to the Theory of Numbers* (Oxford, England: Clarendon Press, 1979), p. 358.

[9]Robert Kanigel, *The Man Who Knew Infinity*, p. 135.

[10]Information of g(n) taken from Paulo Ribenboim, *The Book of Prime Number Records*, pp. 240–245.

## CHAPTER 13

[1]Robert Kanigel, *The Man Who Knew Infinity* (New York: Charles Scribner's Sons, 1991), p. 220.

[2]Paulo Ribenboim, *College Mathematics Journal*, Vol. 25, No. 4, September 1994, 288.

[3]Oystein Ore, *Number Theory and Its History* (New York: Dover Publications, 1976), p. 78.

[4]E.T. Bell, *Men of Mathematics*, p. 486.

[5]E.T. Bell, *Mathematics: Queen and Servant of Science* (New York: McGraw-Hill Book Company, 1951), p. 194.

[6]C. Stanley Ogilvy and John T. Anderson, *Excursions in Number Theory* (New York: Dover Publications, 1966), p. 96.

[7]Paulo Ribenboim, *The Book of Prime Number Records* (New York: Springer-Verlag, 1989), p. 180.

[8]Keith Devlin, *Mathematics: The New Golden Age* (New York: Penguin Books, 1988), p. 210.

[9]D.E. Smith, *History of Mathematics*, Vol. 1 (New York: Dover Publications, 1951), p. 504.

[10]Devlin, *Mathematics: The New Golden Age*, p. 218.

[11]Ribenboim, *The Book of Prime Number Records*, p. 170.

[12]Devlin, *Mathematics: The New Golden Age*, p. 219

[13]Ian Stewart, *The Problems of Mathematics* (New York: Oxford University Press, 1987), p. 126.

[14]James R. Newman, ed., *The World of Mathematics* (New York: Simon and Schuster, 1956), p. 2026.

[15]Devlin, *Mathematics: The New Golden Age*, p. 215.

## CHAPTER 14

[1]John Locke, *An Essay Concerning Human Understanding* (CD: DeskTop BookShop) (Indianapolis: WeMake CDs, Inc., 1994).

[2]J.M. Dubbey, *Development of Modern Mathematics* (New York: Crane, Russak & Company, 1970), p. 112.

[3]Rudy Rucker, *Infinity and the Mind* (New York: Bantam Books, 1982), p. 177.

[4]James R. Newman, "The Foundations of Mathematics," in *The World of Mathematics* (New York: Simon and Schuster, 1956), p. 1616.

[5]Dubbey, *Development of Modern Mathematics*, p. 113.

[6]Rucker, *Infinity and the Mind*, p. 169.

[7]Ernest Nagel and James R. Newman, *Gödel's Proof* (New York: New York University Press, 1958), p. 3.

[8]Most of the following terminology and examples are taken from the Nagel and Newman book, *Gödel's Proof.*

[9]Ibid, p. 68.

[10]Ian Stewart, *The Problems of Mathematics* (New York: Oxford University Press, 1987), p. 218.

[11]John D. Barrow, *Pi in the Sky* (New York: Little, Brown and Company, 1992), p. 122.

[12]Rucker, *Infinity and the Mind,* p. 176.

[13]Barrow, *Pi in the Sky,* p. 117.

[14]Ibid, p. 123.

[15]Rucker, *Infinity and the Mind,* p. 181.

[16]Barrow, *Pi in the Sky,* p. 261.

[17]Douglas R. Hofstadter, *Metamagical Themas: Questing for the Essence of Mind and Pattern* (New York: Basic Books, Inc., 1985).

[18]Kathleen Freemen, *Ancilla to The Pre-Socratic Philosophers* (Cambridge, MA: Harvard University Press, 1966), p. 9.

[19]Hofstadter, *Metamagical Themas,* pp. 11–13.

[20]Ibid, p. 14.

[21]Nagel and Newman, *Gödel's Proof,* p. 97.

[22]Thomas Paine, *Age of Reason* (CD: DeskTop BookShop) (Indianapolis: WeMake CDs, Inc., 1994).

# SUGGESTED READING

Aristotle, *The Basic Works of Aristotle*, ed. Richard McKeon (New York: Random House, 1941).

John D. Barrow, *Pi in the Sky* (New York: Little, Brown and Company, 1992).

Albert H. Beiler, *Recreations in the Theory of Numbers* (New York: Dover Publications, 1966).

Bruce C. Berndt, *Ramanujan's Notebooks, Vol. I & II* (New York: Springer Verlag, 1985).

Eric Temple Bell, *Mathematics: Queen and Servant of Science* (New York: McGraw-Hill Book Company, 1951).

Eric Temple Bell, *Men of Mathematics* (New York: Simon & Schuster, 1965).

Eric Temple Bell, *The Magic of Numbers* (New York: Dover Publications, 1974).

E. J. Borowski and J.M. Borwein, *The HarperCollins Dictionary of Mathematics* (New York: HarperCollins Publishers, 1991).

Carl B. Boyer, *A History of Mathematics* (New York: John Wiley and Sons, 1968).

T.J. Bromwich, *An Introduction to the Theory of Infinite Series* (New York: Chelsea Publishing Company, 1991).

Lucas Bunt, Phillip Jones, and Jack Bedient, *The Historical Roots of Elementary Mathematics* (New York: Dover Publications, 1976).

Florian Cajori, *A History of Elementary Mathematics* (London: MacMillan Company, 1924).

Thomas Crump, *The Anthropology of Numbers* (New York: Cambridge University Press, 1990).

Joseph Warren Dauben, *Georg Cantor: His Mathematics and Philosophy of the Infinite* (Princeton, New Jersey: Princeton University Press, 1979).

Donald M. Davis, *The Nature and Power of Mathematics* (Princeton, New Jersey: Princeton University Press, 1993).

Philip J. Davis, *The Lore of Large Numbers* (New York: Random House, 1961).

Korra Deaver, *The Master Numbers* (Alameda, California: Hunter House, 1993).

Richard Dedekind, *Essays on the Theory of Numbers* (La Salle, Illinois: Open Court Publishing Company, 1948).

Keith Devlin, *Mathematics: The New Golden Age* (London: Penguin Books, 1988).

Heinrich Dörrie, *100 Great Problems of Elementary Mathematics* (New York: Dover Publications, 1965).

J.M. Dubbey, *Development of Modern Mathematics* (New York: Crane, Russak & Company, 1970).

William Dunham, *Journey Through Genius: The Great Theorems of Mathematics* (New York: John Wiley & Sons, 1990).

Euclid, *Elements* (New York: Dover Publications, 1956).

Graham Flegg, *Numbers: Their History and Meaning* (New York: Schochken Books, 1983).

Graham Flegg, *Numbers Through the Ages* (London: MacMillan Education Ltd, 1989).

Kathleen Freeman, *Ancilla to the Pre-Socratic Philosophers* (Cambridge, Massachusetts: Harvard University Press, 1966).

J. Newton Friend, *Numbers: Fun & Facts* (New York: Charles Scribner's Sons, 1954).

J. Newton Friend, *More Numbers: Fun & Facts* (New York: Charles Scribner's Sons, 1961).

Richard Gillings, *Mathematics in the Time of the Pharaohs* (New York: Dover Publications, 1972).

Migene González-Wippler, *The Complete Book of Spells, Ceremonies, and Magic* (New York: Crown Publishers, 1978).

Jacques Hadamard, *The Psychology of Invention in the Mathematical Field* (New York: Dover Publications, 1945).

G.H. Hardy, *A Course of Pure Mathematics* (Cambridge, England: Cambridge University Press, 1963).

G.H. Hardy and E.M. Wright, *An Introduction to the Theory of Numbers* (Oxford: Clarendon Press, 1979).

Sir Thomas Heath, *A History of Greek Mathematics* (London: Oxford University Press, 1921).

Douglas R. Hofstadter, *Metamagical Themas: Questing for the Essence of Mind and Pattern* (New York: Basic Books, Inc, 1985).

Stuart Hollingdale, *Makers of Mathematics* (London: Penguin Books, 1989).

Ross Honsberger, *Mathematical Gems II* (Washington, DC: The Mathematical Association of America, 1976).

Ross Honsberger, *Mathematical Plums* (Washington, DC: The Mathematical Association of America, 1979).

H.E. Huntley, *The Divine Proportion* (New York: Dover Publications, 1970).

A.E. Ingham, *The Distribution of Prime Numbers* (Cambridge, England: Cambridge University Press, 1990).

George Cheverghese Joseph, *The Crest of the Peacock: Non-European Roots of Mathematics* (London: Penguin Books, 1991).

E. Kamke, *Theory of Sets* (New York: Dover Publications, 1950).

Robert Kanigel, *The Man Who Knew Infinity: A Life of the Genius Ramanujan* (New York: Charles Scribner's Sons, 1991).

Jerry P. King, *The Art of Mathematics* (New York: Plenum Publishing Corporation, 1992).

Morris Kline, *Mathematics: A Cultural Approach* (Reading, Massachusetts: Addison-Wesley Publishing Company, 1962).

Morris Kline, *Mathematical Thought from Ancient to Modern Times,* Vol. 1 (New York: Oxford University Press, 1972).

Konard Knopp, *Infinite Sequences and Series* (New York: Dover Publications, 1956).

Konard Knopp, *Theory and Application of Infinite Series* (New York: Dover Publications, 1990).

Stephan Körner, *The Philosophy of Mathematics* (New York: Dover Publications, 1968).

John McLeish, *Number* (New York: Fawcett Columbine, 1991).

Karl Menninger, *Number Words and Number Symbols: A Cultural History of Numbers* (New York: Dover Publications, 1969).

Michael Moffatt, *The Ages of Mathematics,* Vol. 1, *The Origins* (New York: Doubleday & Company, 1977).

Jane Muir, *Of Men and Numbers* (New York: Dodd, Mead & Company, 1961).

Ernest Nagel and James R. Newman, *Gödel's Proof* (New York: New York University Press, 1958).

James Newman, ed., *The World of Mathematics* (New York: Simon and Schuster, 1956).

Carroll Newsom, *Mathematical Discourses* (Englewood Cliffs, New Jersey: Prentice-Hall, 1964).

Ivan Niven, *Numbers: Rational and Irrational* (Washington, DC: The Mathematical Association of America, 1961).

C. Stanley Ogilvy and John T. Anderson, *Excursions in Number Theory* (New York: Dover Publications, 1966).

Oystein Ore, *Number Theory and Its History* (New York: Dover Publications, 1976).

Theoni Pappas, *The Joy of Mathematics* (San Carlos, California: World Wide Publishing/Tetra, 1989).

John Allen Paulos, *Innumeracy: Mathematical Illiteracy and Its Consequences* (New York: Hill and Wang, 1988).

Rozsa Peter, *Playing with Infinity* (New York: Dover Publications, 1961).

Ivars Peterson, *Islands of Truth: A Mathematical Cruise* (New York: W.H. Freeman and Company, 1990).

Plato, *The Dialogues of Plato*, trans. B. Jowett (New York: Random House, 1937).

Earl D. Rainville, *Infinite Series* (New York: The MacMillan Company, 1967).

H.L. Resnikoff and R.O. Wells, Jr., *Mathematics in Civilization* (New York: Dover Publications, 1984).

Paulo Ribenboim, *The Book of Prime Number Records* (New York: Springer-Verlag, 1989).

Gay Robins and Charles Shute, *The Rhind Mathematical Papyrus* (New York: Dover Publications, 1987).

Rudy Rucker, *Infinity and the Mind* (New York: Bantam Books, 1982).

Bertrand Russell, *The Problems of Philosophy* (London: Oxford University Press, 1959).

W. W. Sawyer, *Mathematician's Delight* (London: Penguin Books, 1943).

Annemarie Schimmel, *The Mystery of Numbers* (Oxford: Oxford University Press, 1993).

Denise Schmandt-Besserat, *Before Writing*, Vol. I: *From Counting to Cuneiform* (Austin, Texas: University of Texas Press, 1992).

David Eugene Smith, *History of Mathematics*, Vol. 1 (New York: Dover Publications, 1951).

Sherman K. Stein, *Mathematics: The Man-made Universe* (San Francisco: W. H. Freeman and Company, 1963).

Ian Stewart, *The Problems of Mathematics* (Oxford: Oxford University Press, 1987).

Lloyd Strayhorn, *Numbers and You* (New York: Ballantine Books, 1987).

Dirk J. Struik, *A Concise History of Mathematics* (New York: Dover Publications, 1967).

Frank J. Swetz, *Capitalism and Arithmetic* (La Salle, Illinois: Open Court Publishing, 1987).

Thomas Taylor, *The Theoretic Arithmetic of the Pythagoreans* (New York: Samuel Weiser, 1972).

David Wells, *Curious and Interesting Numbers* (London: Penguin Books, 1986).

W.H. Werkmeister, *A Philosophy of Science* (Lincoln, Nebraska: University of Nebraska Press, 1940).

Alfred North Whitehead, *An Introduction to Mathematics* (New York: Oxford University Press, 1958).

Guy Woodruff and David Premack, "Primate Mathematical Concepts in the Chimpanzee: Proportionality and Numerosity," *Nature* 293 (October 15, 1981).

Claudia Zaslavsky, *Africa Counts* (New York: Lawrence Hill Books, 1973).

Leo Zippin, *Uses of Infinity* (Washington, DC: The Mathematical Association of America, 1962).

# INDEX

one
one one
two ones
one two and one one
one one, one two, two ones
three ones, two twos, one one
one three, one one, two twos, two ones